China-ASEAN Clean Energy Capacity Building Programme

Effective Utilization of Diverse Energy Resources—Development, Application and Practices of Hybrid Energy System

China Renewable Energy Engineering Institute

中国水利水电出版社
China Water & Power Press
· 北京 ·

Abstract

This book, as one of the series publications of China-ASEAN Clean Energy Capacity Building Programme Technical Materials, comprehensively outlines the technical theory, application scenarios and future trends of the utilization of diverse energy resources. Taking the current situation of ASEAN energy development into fully consideration, the book puts forward the needs of developing hybrid energy system with the utilization of diverse energy resources, looks into its prospects in ASEAN region, and explains its application methods with the illustration of comprehensive aspects under abundant practical cases. This book is of significant reference for the theory system of diverse energy hybrid energy system development and a great support to the high-quality development of clean energy.

This book can serve as a reference for practitioners in the fields of clean energy power generation, comprehensive energy planning, international engineering business as well as for teachers and students of relevant majors.

图书在版编目（CIP）数据

中国-东盟清洁能源能力建设计划 ：多能互补技术发展、应用和实践 = China-ASEAN Clean Energy Capacity Building Programme: Effective Utilization of Diverse Energy Resources-Development, Application and Practices of Hybrid Energy System ：英文 / 水电水利规划设计总院编著. -- 北京 ：中国水利水电出版社，2019.9

ISBN 978-7-5170-7949-1

Ⅰ. ①中… Ⅱ. ①水… Ⅲ. ①无污染能源－能源发展－研究－中国、东南亚国家联盟－英文 Ⅳ. ①F426.2 ②F433.62

中国版本图书馆CIP数据核字(2019)第196783号

书　　名	China-ASEAN Clean Energy Capacity Building Programme **Effective Utilization of Diverse Energy Resources** **—Development, Application and Practices of Hybrid Energy System**
中文书名拼音	DUONENG HUBU JISHU FAZHAN、YINGYONG HE SHIJIAN
作　　者	China Renewable Energy Engineering Institute
出版发行	中国水利水电出版社 （北京市海淀区玉渊潭南路1号D座　100038） 网址：www. waterpub. com. cn E-mail：sales@waterpub. com. cn 电话：(010) 68367658（营销中心）
经　　售	北京科水图书销售中心（零售） 电话：(010) 88383994、63202643、68545874 全国各地新华书店和相关出版物销售网点
排　　版	中国水利水电出版社微机排版中心
印　　刷	天津嘉恒印务有限公司
规　　格	184mm×260mm　16开本　9.25印张　301千字
版　　次	2019年9月第1版　2019年9月第1次印刷
印　　数	001—800册
定　　价	**108.00**元

China-ASEAN Clean Energy Capacity Building Programme
Technical Materials

Compilation Working Group

ADVISORY GROUP

Director

LI Fanrong	National Energy Administration of China

Deputy Director

HE Yang	National Energy Administration of China
ZHENG Sheng'an	China Renewable Energy Engineering Institute
Dr. Nuki Agya UTAMA	ASEAN Centre for Energy
HU Jianwu	National Energy Administration of China
GU Hongbin	China Renewable Energy Engineering Institute

Members

YANG Yang	National Energy Administration of China
YAN Bingzhong	China Renewable Energy Engineering Institute
Beni SURYADI	ASEAN Centre for Energy
ZHANG Muzi	China Renewable Energy Engineering Institute
WEI Yingjie	China Renewable Energy Engineering Institute

GUIDANCE

National Energy Administration of China

TECHNICAL SUPPORT

China Renewable Energy Engineering Institute

“中国-东盟清洁能源能力建设计划”系列技术材料

编 写 工 作 组

“中国-东盟清洁能源能力建设计划”工作领导小组

组　长

李凡荣	国家能源局

副组长

何　洋	国家能源局
郑声安	水电水利规划设计总院
Dr. Nuki Agya UTAMA	东盟能源中心
胡建武	国家能源局
顾洪宾	水电水利规划设计总院

成　员

杨　洋	国家能源局
严秉忠	水电水利规划设计总院
Beni SURYADI	东盟能源中心
张木梓	水电水利规划设计总院
魏颖婕	水电水利规划设计总院

指导单位

中国国家能源局

技术负责单位

水电水利规划设计总院

The Compilation Organizations and Group of *Effective Utilization of Diverse Energy Resources—Development, Application and Practices of Hybrid Energy System*

COMPILATION ORGANIZATIONS

China Renewable Energy Engineering Institute

ASEAN Centre for Energy

POWERCHINA Northwest Engineering Corporation Limited

POWERCHINA Huadong Engineering Corporation Limited

POWERCHINA Zhongnan Engineering Corporation Limited

State Grid Xinyuan Zhangjiakou Wind, Solar and Power Storage Demonstration Power Station Co., Ltd.

POWERCHINA Shanghai Electric Power Design Institute Co., Ltd.

Golden Concord Holdings Limited

COMPILATION GROUP

Finalizing

HE Yang	National Energy Administration of China
ZHENG Sheng'an	China Renewable Energy Engineering Institute

Revision

HU Jianwu	National Energy Administration of China
GU Hongbin	China Renewable Energy Engineering Institute
PENG Caide	China Renewable Energy Engineering Institute
LI Pujian	POWERCHINA Northwest Engineering Corporation Limited

Edition

LIU Mingyang	National Energy Administration of China
YAN Bingzhong	China Renewable Energy Engineering Institute

WANG Sheliang   POWERCHINA Northwest Engineering Corporation Limited
MA Li   POWERCHINA Northwest Engineering Corporation Limited
LIU Hanmin   State Grid Xinyuan Zhangjiakou Wind, Solar and Power Storage Demonstration Power Station Co., Ltd.

COMPILATION

YANG Yang   National Energy Administration of China
ZHANG Muzi   China Renewable Energy Engineering Institute
DENG Zhenchen   POWERCHINA Zhongnan Engineering Corporation Limited
WEI Yingjie   China Renewable Energy Engineering Institute
WANG Sheliang   POWERCHINA Northwest Engineering Corporation Limited
YANG Ting   POWERCHINA Northwest Engineering Corporation Limited
HUANG Weiping   POWERCHINA Northwest Engineering Corporation Limited
CHEN Gang   POWERCHINA Northwest Engineering Corporation Limited
XI Yu   POWERCHINA Northwest Engineering Corporation Limited
ZHANG Kai   POWERCHINA Northwest Engineering Corporation Limited
ZHAO Yue   POWERCHINA Northwest Engineering Corporation Limited
HUANG Jieting   China Renewable Energy Engineering Institute
LIU Shujie   POWERCHINA Huadong Engineering Corporation Limited
ZANG Peng   State Grid Xinyuan Zhangjiakou Wind, Solar and Power Storage Demonstration Power Station Co., Ltd.
CHEN Xiaofeng   POWERCHINA Huadong Engineering Corporation Limited
SUN Lei   Golden Concord Holdings Limited
JIANG Zhefan   POWERCHINA Shanghai Electric Power Design Institute Co., Ltd.
REN Yan   China Renewable Energy Engineering Institute
LIU Xueqi   China Renewable Energy Engineering Institute
WANG Yicheng   China Renewable Energy Engineering Institute

TRANSLATION

WEI Yingjie   China Renewable Energy Engineering Institute
QIAO Peng   POWERCHINA Northwest Engineering Corporation Limited
LIU Xueqi   China Renewable Energy Engineering Institute
WANG Yicheng   China Renewable Energy Engineering Institute

《多能互补技术发展、应用和实践》编写单位及成员名单

编 写 单 位

水电水利规划设计总院
东盟能源中心
中国电建集团西北勘测设计研究院有限公司
中国电建集团华东勘测设计研究院有限公司
中国电建集团中南勘测设计研究院有限公司
国网新源张家口风光储示范电站有限公司
中国电建集团上海电力设计院有限公司
协鑫（集团）控股有限公司

编写组成员名单

核 定

何 洋 国家能源局
郑声安 水电水利规划设计总院

审 查

胡建武 国家能源局
顾洪宾 水电水利规划设计总院
彭才德 水电水利规划设计总院
李蒲健 中国电建集团西北勘测设计研究院有限公司

校 核

刘明阳 国家能源局
严秉忠 水电水利规划设计总院
王社亮 中国电建集团西北勘测设计研究院有限公司
马 理 中国电建集团西北勘测设计研究院有限公司

刘汉民　国网新源张家口风光储示范电站有限公司

编　写

杨　洋　国家能源局

张木梓　水电水利规划设计总院

邓振辰　中国电建集团中南勘测设计研究院有限公司

魏颖婕　水电水利规划设计总院

王社亮　中国电建集团西北勘测设计研究院有限公司

杨　婷　中国电建集团西北勘测设计研究院有限公司

黄蔚萍　中国电建集团西北勘测设计研究院有限公司

陈　刚　中国电建集团西北勘测设计研究院有限公司

奚　瑜　中国电建集团西北勘测设计研究院有限公司

章　凯　中国电建集团西北勘测设计研究院有限公司

赵　越　中国电建集团西北勘测设计研究院有限公司

黄洁亭　水电水利规划设计总院

刘树洁　中国电建集团华东勘测设计研究院有限公司

臧　鹏　国网新源张家口风光储示范电站有限公司

陈晓峰　中国电建集团华东勘测设计研究院有限公司

孙　蕾　协鑫（集团）控股有限公司

江哲帆　上海电力设计院有限公司

任　艳　水电水利规划设计总院

刘雪琪　水电水利规划设计总院

王艺澄　水电水利规划设计总院

翻　译

魏颖婕　水电水利规划设计总院

乔　鹏　中国电建集团西北勘测设计研究院有限公司

刘雪琪　水电水利规划设计总院

王艺澄　水电水利规划设计总院

Foreword

China-ASEAN Clean Energy Capacity Building Programme

The world energy scenario is witnessing profound changes, with renewable energy becoming a key sector in global energy development. It is the common mission of the international community to accelerate energy transition and achieve green and low-carbon economy. It has also become one of the priorities for ASEAN countries to transform excellent clean energy endowment into a new economic driver by grasping the historical opportunity of a new round of energy transition. Following the vision of innovative, coordinated, green, open and shared development in recent years, China has gained great achievement in renewable energy by further advancing its energy production and consumption reform, transforming energy development mode, adjusting and optimizing energy structure. The capacity building exchanges jointly conducted by China and ASEAN in clean energy is conducive to seeking complementary advantages and further deepening cooperation in clean energy, thus promoting clean energy development and energy transition as well as regional economic integration.

Against this background, ASEAN and China initiated "China-ASEAN Clean Energy Capacity Building Programme" (the Programme). Based on the framework of the Belt and Road Initiative and taking advantages of the well-functioned dialogue platform of East Asia Summit Clean Energy Forum, the Programme aims to push forward clean energy development and sustainable development in the region, share experience of relevant policies, plans and technology applications, and propel exchanges among talents within clean energy field. With the target of 100 policy and technology backbone in ten years, the Programme plans to develop the capacity of 100 talents in policy and technology for ASEAN countries in various fields, such as pumped storage, wind power, solar energy, nuclear power, hydropower and multi-energy utilization within ten years.

The Programme will be carried out by by China Renewable Energy Engineering Institute and ASEAN Centre for Energy.

Established in 1950, China Renewable Energy Engineering Institute (CREEI) is the only institution dedicated to the management of hydro, wind and solar power technologies in China. As one of the first research and advisory bodies established by the National Energy Administration of China (NEA), CREEI provides comprehensive technical support and services for China's renewable energy industry, including policy research, resources survey and planning, industrial development planning, design review, acceptance of projects, quality supervision, standard formulation, information management, and international cooperation. CREEI, entrusted by NEA, also is in charge of National Research Center for Hydro and Wind Power, National R&D Center for Hydropower Technology Engineering, Station for Renewable Energy Cost, General Station for Renewable Energy Project Quality Supervision and National Center for Renewable Energy Information, and etc.

Founded in January 1999, ASEAN Centre for Energy (ACE) is an independent inter-governmental organization that seeks benefits in energy for the ten ASEAN member states. It is committed to promoting economic development and regional integration within ASEAN area, and building and enhancing multilateral cooperation and coordination in energy field.

According to the Programme, CREEI and ACE will annually and jointly hold exchange project under the Programme in one certain field (pumped storage, wind power, solar energy, nuclear power, hydropower and multi-energy utilization), and invite policy or technology officials from China and ASEAN countries for discussion. The exchange project includes two aspects: policy and technology. The former covers policy framework, comprehensive planning, and management method, while the latter covers resources requirement, equipment application, development and operation, grid-connected management, risk control, environment and society, and etc.

China-ASEAN Clean Energy Capacity Building Programme 2018 Exchange Project

Through elaborate planning and organizing, China and ASEAN successfully organized the China-ASEAN Clean Energy Capacity Building Programme 2018 Exchange Project (the 2018 Exchange Project) in Xi'an, Shaanxi Province on May 7 to 12, 2018. With the theme of "Effective Utilization of Diverse Energy Resources-Development, Application and Practices of Hybrid Energy System", the six-day exchange event gathered more than 20 senior energy officials, scholars and experts from China and ASEAN member states. The participants had extensive and in-depth discussion on the development, application and practice of multi-energy utilization technology. The 2018 Exchange Project realized the profound exchange of clean energy talents between China and ASEAN member states and positively promoted regional energy cooperation.

Effective Utilization of Diverse Energy Resources: Development, Application and Practices of Hybrid Energy System

The modern energy system promotes the utilization of clean, low-carbon and efficient energy production and supply utilities. It is oriented by the demand and integrates diverse energy types with coordinated supply, shared development, and smart interaction between the supply and demand sides. The transition from the conventional energy system to the modern energy system requires fundamental renovation of the energy structure, supply mode and supply/demand relation. Hybrid energy system is a necessary path to realize the concept of modern energy system. Through system integration, strengths enhancement and structure optimization, it captures the key aspects of achieving lower carbon emission, cleaner energy, higher efficiency, and lower cost, and by synthetic integration of diverse energy sources and coupled supply featuring clean energy propels, it also propels the the interaction between supply and demand sides. As a novel method of energy production and utilization, hybrid energy system can effectively resolve the difficulties facing clean renewable energy production, transmission and consumption. The research and practice over the years demonstrate that sound application of hybrid energy system with the power sources, such as hydropower, coal-fired power, wind power, and solar power, can effectively promote the consumption of clean renewable energy, and thus improve electric energy quality, transmission efficiency, energy utilization efficiency and operation cost-efficiency. At the consumption side, the application of hybrid energy system in smart micro-grid can, to a greater extent, realize energy access in the off-grid regions and mitigate the electricity access deficits. Therefore, hybrid energy system plays a significant role in the transition process from conventional energy system to modern energy system.

During the 2018 Exchange Project, China Renewable Energy Engineering Institute (CREEI), ASEAN Center for Energy (ACE), POWERCHINA Northwest Engineering Corporation Limited, POWERCHINA Huadong Engineering Corporation Limited, State Grid Xinyuan Zhangjiakou Wind, Solar and Power Storage Demonstration Power Station, POWERCHINA Shanghai Electric Power Design Institute Co., Ltd and GCL Design & Research Institute jointly issued the technical materials for China-ASEAN Clean Energy Capacity Building Programme 2018 Exchange Project, which received high praises and positive responses of the officials and enterprise representatives participated in the event.

Effective Utilization of Diverse Energy Resources: Development, Application and Practices of Hybrid Energy System is the book based on the technical material for the 2018 Exchange project. It covers all the aspects relevant to the development, application and practice of hybrid energy system. It was developed through deep cooperation between Chinese research institutes and engineering enterprises and international organizations. During the process of planning and compilation, the compiling group had conducted in-depth investigation of ASEAN's clean energy demand and endeavored to provide effective and concrete reference and advice to the cooperation and development of hybrid energy system in China and ASEAN.

As the second work of the serial publication of China-ASEAN Clean Energy Capacity Building Programme, the *Effective Utilization of Diverse*

Energy Resources: Development, Application and Practices of Hybrid Energy System will lay a solid foundation of positive cooperation and joint development in the clean energy sectors of both China and ASEAN.

Compiler

February 2019

编者的话

中国-东盟清洁能源能力建设计划

当今世界能源形势正在发生深刻的变化，可再生能源已成为国际能源发展的重要领域，加快能源转型，实现绿色低碳发展，已经成为国际社会的共同使命。如何抓住新一轮能源变革的历史机遇，将优越的清洁资源禀赋转化为经济发展的新动力，已经成为东盟国家关注的重点之一。近年来，中国秉持“创新、协调、绿色、开放、共享”五大发展理念，深入推进能源生产消费革命，转变能源发展方式，调整优化能源结构，可再生能源取得飞跃式发展。中国与东盟国家共同开展清洁能源能力建设交流，有助于实现优势互补，进一步深化双方在清洁能源领域的合作，共同促进清洁能源发展和能源转型，推动区域经济一体化进程。

正是在此背景下，中国和东盟共同开展“中国-东盟清洁能源能力建设计划”（以下简称“能力建设计划”）。能力建设计划基于“一带一路”的愿景框架，借力东亚峰会清洁能源论坛的良好对话平台，旨在推动区域清洁能源和可持续发展，分享清洁能源发展政策规划和技术应用等经验，推进相关领域的核心人才交流建设。能力建设计划以“十年百位政策技术骨干”为目标，针对抽水蓄能、风电、太阳能、核电、传统水电、多能互补等专题领域，计划在10年间共同为东盟国家培养百位政策技术骨干。

能力建设计划由水电水利规划设计总院和东盟能源中心共同实施。

水电水利规划设计总院（以下简称“水电总院”）成立于1950年，是中国唯一的水电、风电和光伏发电技术归口管理单位。水电总院为国家可再生能源行业政策研究，资源普查与规划，产业发展规划，工程设计审查、验收、质量监督，技术标准制定，信息管理，以及国际合作等方面提供全方位的技术支持和服务，并受国家能源局委托管理国家水能风能研究中心、国家水电技术工程研发中心、可再生能源定额站、水电工程质量监督总站、国家可再生能源信息管理中心等，是国家能源局设立的首批研究咨询基地。

东盟能源中心（ASEAN Centre for Energy，ACE）成立于1999年1月，是独立代表东盟十国能源领域利益的政府间国际组织。ACE致力于推动东盟区域经济发展和区域一体化进程、建立和促进多边合作以及能源领域的协同活动。

根据计划，水电水利规划设计总院和东盟能源中心将每年共同组织一期清洁能源能力建设交流项目，针对一个专题领域（抽水蓄能、风电、太阳能、核电、传统水电、多能互补等），邀请中国和东盟国家的政策或技术官员进行研讨。交流内容分为政策和技术两方面，政策方面包括政策框架、综合规划思路、产业管理手段等；技术方面包括资源要求、设备应用、开发运维、并网管理、风险控制、环境社会等。

中国-东盟清洁能源能力建设计划2018交流项目

经过精心策划和积极组织，中国与东盟于2018年5月7—12日在陕西省西安市成功举办了“中国-东盟清洁能源能力建设计划2018交流项目”（以下简称“2018交流项目”）。本次交流主题为“多能互补技术发展、应用和实践”，来自中国和东盟十国的20余位能源主管部门领导、专家、学者在为期6天的交流活动中，共同就多能互补技术的发展、应用和实践进行了广泛深入的研讨。2018交流项目实现了中国与东盟国家清洁能源领域人才的深入交流，对推动区域能源合作起到积极作用。

《多能互补技术发展、应用和实践》

现代能源体系倡导以建立清洁、低碳、高效能源生产供给体系为目标，以需求为主导，多品种能源融合、多种供能方式协同、多元主体开发共享、供需智慧互动的能源系统，从传统能源体系到现代能源体系的转变，需要进行能源结构、能源供应方式、能源供需关系的根本性变革。多能互补正是构建现代能源体系的应有之意和实现途径，多能互补从系统集成、优势互补、结构优化的角度抓住了降低碳排放、实现清洁化、提高能效和降低成本的关键点，通过以清洁能源为主的多种能源的有机整合、多种产品的耦合供应，促成能源供给侧和消费侧的互动。多能互补作为一种新的能源生产利用方式，能够有效地解决清洁可再生能源在生产、传输、使用过程中遇到的问题。近年来在能源领域的探索和实践表明，在电源侧科学合理地应用水火风光等类型电源进行多能互补，能够有效地促进清洁可再生能源的消纳，改善电能质量，提高输电效率，提高能源使用效率，改善运行经济性等；在消费侧，多能互补在智能微电网方面的应用，能够更大程度地实现离网区域的能源可及，减少无电、缺电人口。因此，多能互补是传统能源体系向现代能源体系过渡的重要措施。

2018 交流项目期间，水电水利规划设计总院、东盟能源中心、中国电建集团西北勘测设计研究院有限公司、中国电建集团华东勘测设计研究院有限公司、国网新源张家口风光储示范电站有限公司、中国电建集团上海电力设计院有限公司、协鑫集团设计研究总院发布了“中国-东盟清洁能源能力建设计划”2018 技术交流材料——《多能互补技术发展、应用和实践》，获得了中国和东盟国家政府、企业的高度评价和积极响应。

《多能互补技术发展、应用和实践》是针对 2018 交流项目特别编制的一本技术交流材料，内容涵盖了多能互补技术发展、应用和实践的各个方面。《多能互补技术发展、应用和实践》是由我国科研机构和工程企业与国际组织深入合作完成的技术交流材料。在其策划和编写的过程中，编写组对东盟的清洁能源需求进行了深入调研，力求为中国和东盟多能互补合作发展提供有效、务实的参考和建议。

作为“中国-东盟清洁能源能力建设计划”系列技术交流材料的第二本著作，《多能互补技术发展、应用和实践》将为中国和东盟在清洁能源领域发挥各自优势、协力合作、共同发展打下良好基础。

编者

2018 年 12 月

内 容 提 要

本书作为“中国-东盟清洁能源能力建设计划”系列技术材料之一，全面梳理了多能互补的技术理论、应用场景、未来趋势。全书紧密结合东盟能源发展现状，提出了多能互补的发展需求，展望了东盟多能互补的发展前景，并结合丰富的实践案例，全方位、多角度地阐释了多能互补的应用方式。本书对构建多能互补理论体系、促进清洁能源的高质量发展具有借鉴意义。

本书可供清洁能源发电、综合能源规划、国际工程商务等领域的从业人员阅读，也可供高等院校相关专业师生参考。

Contents

Foreword

Chapter 1
Hybrid Energy System Development in China

Chapter 2
Potentials of Development of Hybrid Energy Systems in ASEAN Region

Chapter 3
Application of Hybrid Energy System

Chapter 4
Practices of Hybrid Energy System

目录

第3章 多能互补的应用

第4章 多能互补实践

Chapter 1

Hybrid Energy System Development in China

Hybrid energy system (also equivalently termed as multi-energy mix system) refers to an energy use mode, in which multiple energy are complemented with each other and effectively utilized in consideration of their unique characteristics and diverse energy consumers, so as to ease the contradictory between energy supply and demand, protect and make use of natural resources rationally, and achieve environmental benefits. In the energy field, hybrid energy system is in the forefront. China and some other countries have carried out researches and promoted pilot applications of hybrid systems, and certain achievements have been made.

1.1 Hybrid Energy System Development

Based on relevant literature, the worldwide (excluding China) hybrid energy system development is briefed as below:

In the 20th century, the theory of hybrid energy system focused on theoretical researches and was in its preliminary stage. In 1981, N. E. Busch et al. from Denmark put forward technical questions for the mixed use of solar and wind energy; C. l. Aspliden from the USA studied meteorological matters related to solar-wind hybrid energy conversion system; scholars from former Soviet Union estimated the approximate potential values of solar and wind energy. In 1996, J. G. McGowan et al. from the USA studied the microgrid including wind power, diesel and other energy resources. In 1997, Jun Hasegawa et al. from Japan proposed the concept of adaptive and distributed microgrid power generation system.

Since the 20th century, the theoretical researches of hybrid energy system have been gradually deepened. Project demonstration was commissioned at this stage. As for the wind-solar hybrid system, Rajesh Karki et al. from Canada studied the cost and reliability of an independent small wind-solar power generation system and Roy Billinton et al. studied the capacity expansion of such system. Rodolfo Dufo Lopez et al. from Spain developed an optimization system for wind-solar and solar/diesel power generation systems; Colorado State University and National Renewable Energy Laboratory jointly developed the application software hybrid 2 which can accurately simulate a wind-solar hybrid system; B. D. Shakva et al. from Australia designed a wind-solar hybrid power generation system equipped with a compressed-hydrogen energy storage plant. As for the hydro-wind hybrid power system and the pumped storage-wind hybrid system, Brook D et al. assessed the coordinated operation of wind power and hydropower for the Pacific Northwest of the United States, Quebec Hydropower System in Canada and Nord Pool, and expounded advantages of a wind power and hydropower hybrid system. Garcia-Gonzalez and Tsunb Ying Lee et al. optimized the model of a coordinated operation system for combining wind power and pumped storage power plants. As for microgrid, Professor R. H. Lasseter proposed the concept of microgrid in 2001. Soon afterwards, some research institutions in the USA and EU successively defined the microgrid, and scholars worldwide furthered their searches on microgrid. Koichi Hidese et al. studied the coordination and optimization of a multi-energy distributed system; Mallikarjuna R. Vallem studied relevant matters of the system controller for microgrid. With the advance of researches on hybrid energy system, some demonstrative practices were gradually being carried out. For example, the University of Wisconsin built a small microgrid laboratory in 2003, with the total capacity of 80 kVA. Greece, Germany and Spain built experimental platforms of different-sized microgrids, among which the microgrid laboratory built by ISET is the largest one with the capacity up to 200 kVA. EI Hierro Island and Canary Island of Spain witnessed the joint operation of pumped storage power plants and wind power plants. UK developed the building combined heat and power (BCHP) for hospitals, hotels, schools and government.

The hybrid energy technology leads the new trend of energy use. From existing data, the theoretical researches worldwide (excluding China) are deepening and project demonstration is at the launch stage. At present, research and application of hybrid energy system witnesses vigorous development in China. The hybrid energy system technology has developed into a relatively complete theoretical system, entered the scale development stage, and several projects have been implemented. China has made great achievements in theoretical research of hybrid energy system, and the application practices of China are representative and leading in the world.

1.2 Hybrid Energy Development in China

1.2.1 Background and Development History

1.2.1.1 Background

Energy is an important material basis for socio-economic development. In this period of profound readjustment of the global economy, a new round of energy revolution is poised to take off. Under the world context of peace and development, China's economy

has maintained sound growth. But the needs for quality and efficiency improvement as well as transformation and upgrading are more urgent, and energy development faces numerous challenges.

The new energy consumption becomes an outstanding issue. Recently, China has actively pushed forward the clean energy substitution to spare fossil energy consumption. In 2017, China's installed capacity of clean energy accounted for 38% of total installed power capacity. The role of clean energy in promoting energy restructuring has continued to grow. China witnessed a rapid development of wind power and PV power generation. In 2017, the installed capacity of wind power and PV power reached 164 GW and 130 GW respectively. Wind power and PV power generation are characterized by fluctuating, intermittent and stochastic. Because of the insufficient peak load regulation, and an incomplete cost compensation system for scheduling operation and peaking operation, the power system can hardly meet the large scale grid-connection and consumption demands of wind power and solar power, which results in the serious problem of abandoning and rationing wind (solar) power supply onto the grid.

The overall efficiency of the energy system is relatively low. Different energy supply systems, including power, heat and gas see relatively low level of integration, mutual complementation and cascade use. The demand-side response mechanism has not been fully established; energy supply cannot be well adapted to the seasonality and fluctuation of energy consumption, and the utilization efficiency of system equipment decreases. In order to solve the problem of unbalanced distribution of the energy resources and loads, the vigorous development of trans-provincial power transmission channels makes great contribution to the energy restructuring and socio-economic development. However, the existing trans-provincial electricity transaction and compensation mechanism and other supportive policies require further improvement. In remote areas such as some islands and high altitude Tibetan region, it's urgent to solve the power supply problem.

Under above context, vigorously implementing the hybrid energy system is favorable to promoting new energy consumption, enhancing the skills of coordinating energy supply and demand, raising the energy use efficiency, and furthering the construction of a clean, low-carbon, safe and efficient modern energy system.

1.2.1.2 Development History

In the 20th century, research on hybrid energy system focused on theories and was in the preliminary stage. In 1982, Yu Huayang et al. studied the energy conversion device for solar and wind-driven generators; in 1997, Bao Xiaoqing et al. studied the complementary operation of hydropower and wind power. In 1994, Burqin County, Altay Prefecture, Xinjiang carried out practical exploration in the aspect of hydro-wind power complementary operation while conducting theoretical studies.

From 2000 to 2010, the theoretical researches on hybrid energy system had been gradually deepened. Project demonstration was at the launch stage. As for wind-solar hybrid system, the Hong Kong Polytechnic University et al. proposed the method of using CAD to optimize the wind-solar hybrid power generation system; Hefei University of Technology put forward the variable structure simulation model of wind-solar power generation system; Shanghai Jiao Tong University investigated the use of household renewable energy power generation systems in Sonid Right Banner, XilinGol League, Inner Mongolia, and made relevant suggestions; Chinese Academy of Agricultural Mechanization Sciences, etc. studied the typical configuration of wind-solar hybrid system for mobile communication, to solve the problem of power supply for equipment in the mobile communication construction. As for the microgrid, Sun Ke et al. studied the application of a micro gas-turbine system in the distributed power generation; Dong Xueping et al., researched and developed a wind-solar-diesel-storage hybrid energy generation and its intelligent control system; Wang Chunming et al. studied and designed a wind-solar-diesel hybrid power supply system; China Southern Power Grid undertook the project of "Research on and Project Demonstration of Key Technologies for Grid Connection of Megawatt CCHP Distributed Energy Microgrid" under the 863 Program; Tianjin University carried out a national major fundamental research plan, i. e. "Fundamental Research on Distributed Power Generation and Energy Supply System", and key research on the distributed power generation microgrid together with Huazhong University of Science and Technology and Xi'an Jiaotong University, etc. Institute of Electrical Engineering of the Chinese Academy of Sciences engaged in the research on the application of microgrid power generation and energy storage technology. In the aspect of hydro-wind hybrid power system and the pumped storage-wind hybrid power system, experts and scholars presented the combined scheduling scheme of wind power and pumped storage power plant and carried out relevant research and exploration. With gradually intensive researches, the construction of multi-energy hybrid system demonstration projects kicked off. In 2000, Yangtze Source Natural Conservation Station was equipped with an independent operational wind-solar hybrid power generation system (capacity: 1,000 W/400 Wp); in 2004, Huaneng Nan'ao 54 MW/100 kWp wind-solar hybrid power plant was successfully connected to the grid and became the first commercially-commissioned wind-solar hybrid power system in China. In 2009, a 50 kW off-grid wind-solar hybrid power plant was built in Naqu Prefecture in Tibet.

Since 2010, theoretical researches on hybrid energy system have witnessed systematical development and project practices developed into large-scale development. The theoretical research involves the hydropower and new energy hybrid system, thermal power and new energy hybrid system, pumped storage and new energy hybrid system, microgrid, and multi-energy integration and optimization. China Renewable Energy Engineering Institute, POWERCHINA Northwest Engineering Corporation Limited et al.

studied effective utilization of diverse energy resources in the northwestern region of China, and established a relatively systematical and improved theoretical system and research methods for hybrid energy system, which shows significance to theoretical and practical guidance. Hybrid energy projects are thriving. In 2013, Xinhua Hetian Bobona 20 MWp Hydro-solar Hybrid Power Plant in Xinjiang was commissioned and connected to the grid; in 2014, Hami-Zhengzhou UHVDC project, the first UHV wind-solar-thermal bundled power transmission project in China was put into operation; in 2015, 850 MWp Longyangxia Hydro-solar Hybrid Power Project, the largest hydro-solar hybrid power project in the world, was connected to the grid; from 2011 to 2015, China built about 50 microgrid projects; in 2016, Shuanghu County renewable energy grid in Tibet, the micgrogrid with the highest altitude in the world, came stream; in 2017, China also determined to build 23 multi-energy integration and optimization demonstration projects and 28 new energy microgrid demonstration projects.

1.2.2 Classification of Hybrid Energy Systems

Hybrid energy systems can be divided into two categories.

1.2.2.1 Hybrid Energy Systems at Source Side

Through the joint complementary operation of multiple power sources in the electric power system including hydropower, wind power, PV and pumped storage, etc., their respective advantages can be fully utilized to meet the power demand, ensure safe and stable operation of grid, and facilitate renewable energy development and consumption while getting better environmental benefit.

Hybrid energy systems at source side include a variety of hybrid forms, including hydropower-wind (solar) power hybrid system, thermal power-wind (solar) power hybrid system, pumped storage-wind (solar) power hybrid system, pumped storage-wind (solar) power hybrid system, natural hybrid of new energy, wind-solar-hydro-thermal-storage hybrid energy system.

1.2.2.2 Hybrid Energy Systems at User Side

In view of power, heating, cooling and air supply demands of end users, the coordinated and complementary development of conventional and new energy should be carried out to optimize the layout of integrated energy supply infrastructures, and realize coordinated supply of multiple energy and comprehensive cascade use of energy through cogeneration of natural gas, heating and cooling, distributed renewable energy, smart microgrid, etc.

1.3 Significance

With intensified environmental and climate change, the traditional energy system has shown the defects of high pollution, high carbon emission and low energy efficiency. As a major solution to the crisis, it's imminent to build a modern energy system. Modern

energy system is designed to be a demand-oriented energy system featuring integration of multiple energy resources, coordination of multiple energy supply methods, joint development and sharing of multiple subjects and intelligent interaction of supply and demand and target to build a clean, low-carbon and efficient energy production and supply system. The transformation from traditional energy system to modern energy system requires fundamental changes of energy structure, energy supply method and energy supply and demand relation. Hybrid energy system is the theme and way of building a modern energy system. It captures the key issues, lower carbon emission, cleanness realization, efficiency improvement and lower cost from the perspective of system integration, advantage complementation and structural optimization, and promote the interaction between energy supply side and consumption side through integration of multiple energy resources (dominated by clean energy) and coupling supply of multiple products. Hybrid energy system, a new energy production and utilization mode, is able to effectively solve problems of clean and renewable energy resources in production, transmission and use. Exploration and practices in the energy field in recent years have shown that the scientific and reasonable application of the hydropower, thermal power, wind power and solar power and others based on hybrid energy systems at the power source side could effectively promote the consumption of clean and renewable energy resources, improve power quality, enhance transmission efficiency and energy utilization efficiency, and improve the operating economy. At the user side, the traditional energy and new energy are developed and utilized in a coordinated and complementary way and an integrated energy supply infrastructure is built to meet various demands of end users on the consumption of electricity, heating and cooling, and realize the coordinated supply of multiple energy resources and the comprehensive cascade use of energy. The hybrid energy application in a smart grid can realize the energy accessibility of off-grid areas and realize the population without access to electricity and short of electricity to a large extent. Therefore, hybrid energy technology is an important measure for transition from the traditional energy system to a modern energy system.

1.3.1 Promoting the Development of Clean Energy

Clean energy is the trend of future energy development, but the development of clean energy is still facing a series of difficulties. For example, large-scale integration of new energy sources will threaten the safe and stable operation of the power system; new energy features relatively high cost and poor economy; structural problems in the process of large-scale development, power curtailment occurs frequently. Hybrid energy technology provides a new solution to the above issues. Through the compensative regulation of a variety of power supplies in a scientific and reasonable way we can improve the power quality of new energy, improve operational efficiency, reduce power curtailment and improve power economy, promote the development of clean and renewable energy, and achieve low-carbon, clean and efficient energy use.

1.3.1.1 Improving the Power Quality of New Energy and Stabilizing Output Fluctuation

The power system is characterized by continuity and simultaneity. The power generation, transmission and utilization are completed simultaneously and continuously, and the power supply adjusts the power generation and output according to the change of the power load to keep the grid voltage and frequency stable. But the output of wind power and solar power features intermittency, randomness and volatility. They have no self-regulating capability, and their power quality is poor. Thus, large-scale grid interconnection will impact the power system. Through a hybrid energy system, wind power and solar power are coupled with power sources of certain regulation capability and rapid response (such as hydropower, pumped storage and gas turbines) for joint operation. When the output of new energy is excessive, the hydropower plant should store water (or pumped storage power plant should pump water by using extra electrical energy). When the output of new energy decreases, the hydropower should make up the output of new energy, so as to stabilize the fluctuation of new energy output when operating alone, improve its power quality and reduce the impact of its large-scale grid connection on the system.

1.3.1.2 Improving Operating Condition of Clean Energy and Reducing Power Curtailment

Due to the poor power regulation of wind power, solar power and others, their output and load do not match with each other. When the load is great at night, the output of solar power is reduced, and similarly wind energy is also volatile and intermittent. For regional power grid with relatively high proportion of new energy, if new energy is forced to be involved in peak load regulation, large-scale power curtailment will occur. Through a hybrid system, wind power and solar power are coupled with certain scale peaking power supply (such as regulating hydropower, pumped storage and gas turbine) for joint operation, which will improve operating conditions of new energy, give full play to the power efficiency and reduce the curtailment loss of new energy.

1.3.1.3 Enhancing the Utilization Efficiency of Energy Transmission Channel

To further the large-scale development and utilization of clean energy resources, China has planned large-scale clean energy bases such as the "Northwest, North China and Northeast" wind power bases and the Northwest Solar Power Generation Base. Due to the reverse distribution of load center and clean energy base in the geographical location, the long-distance transmission of clean energy has become the key factor for clean energy consumption. Taking wind power and PV as an example, in the case of large-scale and long-distance transmission and grid connection, low hours of wind power and PV power generation, the large scale but low utilization hours of transmission channels will harm the transmission line economy and further adversely affect the construction of the base. With a hybrid energy system, bundling up other power supplies (including hydropower, pumped storage, gas power and thermal power) upon the power transmission of clean energy base, and increasing the output of other power supplies and the utilization hours of power transmission channels upon the low output of wind power and PV power, could effectively increase the power of clean energy to be delivered, improve the overall economy and encourage the construction of clean energy base.

1.3.2 Improving Local Energy Supply System and Promoting Integrated, Intelligent and Efficient Application of Energy

Another way of effective utilization of diverse energy resources is the terminal integrated energy supply system. This model is mainly used in terminals. For new towns, new industrial parks, large-scale public facilities, and other infrastructures, this model realizes the joint development and utilization of traditional energy and wind energy, solar energy, geothermal energy, biomass energy and others, optimizes the configuration of such infrastructures as power, gas, heating, cooling and water supply pipe gallery, provides users with efficient and intelligent energy supply and related value-added services, promotes in-situ production and consumption of clean energy, closely combines production with consumption. The energy scheduling platform is built to conduct comprehensive dispatch and control of energy input and output, promote the intelligent development of energy and enhance the comprehensive utilization efficiency of energy. The terminal integrated energy supply system is also the key direction for the development of hybrid energy systems.

1.3.3 Enhancing Energy Accessibility and Provide Electricity for Areas without Access to Electricity

By 2016, the global population without access to electricity was still up to 1.1 billion, accounting for 14% of the world's population. According to the current trend, the IEA forecasts that there will be still 9% of the world's population that cannot get access to electricity by 2030. In rural areas, remote areas and islands of South Asia, Southeast Asia and Pacific Island countries and regions, the energy shortage is particularly serious.

Smart microgrid, an application form of hybrid energy systems, could solve the power supply problem of off-grid densely populated areas on islands or in high-altitude and severe cold regions. Take a sea island for example. Local energy supply to smart microgrids consists of large quantity of distributed renewable energy, (such as wind energy, solar energy and wave energy) whose their output power has such shortcomings as intermittency, volatility and lack of scheduling. To solve the problem, smart microgrids integrate multiple energy input ways such as renewable energy and energy storage devices. Based on integrated and intelligent scheduling, a smart microgrid can realize multi-energy output such as electrical energy, heating (cooling) and water, and provide a new solution to the energy shortage in off-grid regions. The utilization of smart microgrid plays an active role in improving energy accessibility and reducing the population without access to electricity.

1.3.4 Improving Energy Infrastructure to Boost Local Economy

To build hybrid energy projects and improve energy infrastructures and establish a clean, low-carbon, green and efficient modern energy system is an important approach to make renewable energy sources account for 23% of total primary energy consumption by 2025 as set by ASEAN. Such projects will attract enormous social investment, effectively promote the infrastructure construction at the localities of these projects, and boost the GDP growth as well as the development of related industries in the surrounding areas. In addition, hybrid energy technology encourages clean energy industries and effectively increases local employment. Calculated according to IRENA's 2016 employed population and proportions of clean energy power installed capacity, the construction of 1 MW clean energy could create jobs for 5 people. In addition, the PV-based poverty alleviation projects in China have produced steady earnings for 25 consecutive years by fully utilizing and developing solar energy resources in the poverty-stricken areas. In this way, China combines poverty alleviation with new energy utilization, energy saving and emission reduction. China provides, through PV power generation projects, long-term compensation to underdeveloped areas for targeted poverty alleviation.

1.4 Application Prospect

1.4.1 Development Foundation

1.4.1.1 Rapid Development of Renewable Energy Sources

Nowadays, the world energy structure is undergoing profound adjustment. Against global warming, vigorously developing renewable energy to replace fossil energy has become an irresistible trend for energy transition in multiple countries. In recent years, we see the rapid development and utilization of global solar and wind energy, continuous technical progress and significant cost reduction, marking them promising. Particularly, the entire PV power generation industry is undergoing large-scale development. The traditional PV power markets such as China, Europe, America and Japan keep on rapid growth and the emerging PV power markets such as Southeast Asia, Latin America, Middle East and Africa also develop rapidly. In 2016, the accumulative PV power generation worldwide reached 303 GW from 1.3 GW in 2000, registering an annual growth rate of 41%. In 2017, the newly added capacity reached 100 GW for the first time, making PV the fastest-growing energy resource in the world. It's predictable that PV, wind power and other renewable energy sources will lead the global energy development.

1.4.1.2 Gradual Maturation of Energy Storage Technology

Energy storage is an important way of energy management of the power system. With the accelerating development and utilization of new energy sources, the energy storage technology will embrace rapid development to safeguard the security and stability of the grid. In the future, energy storage will take up an increasingly important position in the energy sector. The preferred technologies of energy storage include pumped storage and chemical energy storage. According to the demonstration studies, to stabilize power supply and provide uniform power output, a battery energy storage system with capacity being 20% of new energy generating capacity and storage time being 6-8 hours is required. It's expected that, a total of 758 GW energy storage devices will be required at the power generation and consumption sides by 2020. It's anticipated that, the installed capacity of global energy storage system will approximately reach 45 GW/81 GWh by 2024. The future power system will embrace qualitative changes along with the development of energy storage system, which will provide a solid basis for large-scale application of hybrid energy system.

1.4.1.3 Development of Energy Internet

The future energy system shall be characterized by renewableness, cleanness, low carbon and intelligence. To meet these requirements, the concept of Energy Internet emerges. Energy Internet is a network where the new power network nodes composed of distributed energy harvesting devices and storage devices as well as various loads are interconnected by comprehensively utilizing advanced electric and electronic technologies, information technology and intelligent management technology so as to realize the peer exchange and sharing of energy featuring bidirectional flow. The compatibility of Energy Internet with traditional power grids enables the full, wide and effective utilization of distributed renewable energy to meet the diverse power demands of users. At present, most developed economies are actively planning the strategic configuration of Energy Internet. On March 2, 2015, ISO/IEC released official documents, identifying IEEE1888 as the first ISO/IEC standard for global energy Internet industry. Energy Internet is an important in-depth practice of hybrid energy system.

1.4.2 Future Application

Along with the development of renewable energy and energy storage, the energy production and consumption are becoming diversified. The developing Energy Internet will effectively combine multiple energy sources. "Hybrid energy system" will become a new trend of energy sustainable development and lead the energy sector to integrate in-depth the multiple energy resources.

As the load centers and clean energy resources are distributed at different geographical locations in China, the long-distance transmission of clean electric energy has become a key restriction on its growth. An important form of application of hybrid energy system

at the power source side in future is as follows: building a "wind, solar, hydrothermal and energy storage" hybrid system to complement and regulate various energy resources and minimize the output fluctuation and power curtailment; increasing the utilization efficiency of new energy and the utilization rate of the transmission channel and improving the system economy; and facilitating the large-scale trans-regional deployment of renewable energy to promote the establishment of a clean, low-carbon and efficient energy system in the future.

At the user side, an important form of application of effective utilization of diverse energy resources in future is to build a smart city driven by smart energy system. With a hybrid energy system, the input and output of multiple energy resources are coordinated and dispatched intelligently and rationally so as to build a green, low carbon, efficient, intelligent and safe smart energy system (as shown in Figure 1.4.1). Hybrid energy system will be continuously applied in the smart microgrid to sustain self energy supply based on rational interconnection of multiple energy production systems, and the distributed energy production mode at the user side involving hybrid systems of "distributed PV + energy storage", "wind-solar-energy storage", etc., will undergo sustainable development in future. Taking representative examples of addressing energy supply problems in remote areas and on islands, smart microgrids will play a key role in achieving energy accessibility in future.

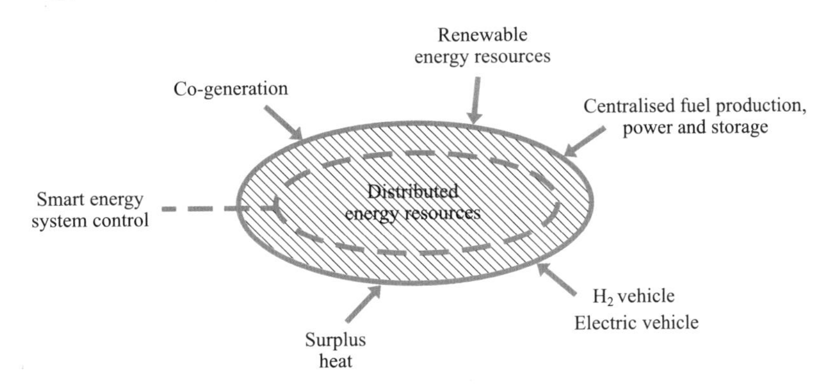

Figure 1.4.1 Smart City Energy System

In summary, with the diversified energy development in the future, the "hybrid energy system" will take off and provide a solution to the energy problems both at the power source side and at the user side in future.

1.5 China's Policy Support for Hybrid Energy System

In the 20th century, China's researches on hybrid energy system focused on theories and were at the preliminary stage. From 2000 to 2010, the theoretical researches on hybrid energy system had been gradually deepened. Demonstration projects were launched one by one. Since 2010, researches on the theories of hybrid energy system have been witnessing systematical development and projects are carried out in large scale. The Chinese government has proactively given many supportive policies on hybrid energy system and issued a series of guidelines. The main policy documents are as follows:

1.5.1 Effective Utilization of Diverse Energy Resources in Planning

1.5.1.1 National Macro Planning

According to *the 13th Five-Year Plan (FYP) for Economic and Social Development of the People's Republic of China*, China should accelerate the construction of a modern energy storage and transportation network that is multi-energy complementary, open, secure and reliable; work faster to provide smart energy in all fields and links; accelerate the construction of smart grids and enhance the interactive response capacity between the grids and the generation and demand sides; promote the in-depth integration of new technologies in energy and information industries and build an energy Internet featuring coordinated development, integration and complementation of "source-grid-load-storage".

1.5.1.2 Energy and Electric Power Development Planning

According to *the 13th FYP Development Plan for Renewable Energy*, to implement integrated and optimized hybrid power projects is one of the main tasks of energy sector during the 13th FYP period. China should strengthen the overall planning and build an integrated terminal energy supply system. In new energy consumption areas, China should implement terminal integrated energy supply projects, apply energy supply modes such as combined cooling, heating and power, distributed renewable energy generation and geothermal heating and cooling according to local conditions, and strengthen the production, coupling, integration and complementary use of heating, electricity, cooling and gas sources. In existing energy-consuming areas, China should promote the integrated cascade utilization and transformation of energy, popularize and apply the above-mentioned energy supply modes, and strengthen the recovery and comprehensive utilization of energy resources such as residual heat and pressure, industrial by-products and domestic wastes. China should give full play to the advantages of wind energy, solar energy, hydropower, coal, natural gas and other resources in large-scale integrated energy bases to promote the construction and operation of the multi-energy hybrid projects of wind, solar, hydro and thermal storage.

According to *the 13th FYP Development Plan for the Electricity Sector* (2016-2020), China should apply and demonstrate in an integrated way diverse wind and solar storage and transmission technologies. In combination with the development of wind power, PV and other new energy and energy storage and microgrids, China should promote the demonstration and diversified application of renewable energy and energy

storage, intelligent power transmission and other new technologies as well as integrated development of new energy power featuring multi-energy complementation, coordination and optimization.

1.5.1.3 Planning for Industrial Distribution

According to *A Guideline on Emerging Sectors of Strategic Importance during the 13th Five-Year Plan Period*, we should work faster to construct the power system mechanism, new power grids and innovation support system that adapt to the rapid development of new energy resources and promote effective utilization and collaborative optimization of diverse energy resources. We should work actively to promote the comprehensive utilization of diverse new energy, accelerate the development of distributed energy sources that integrate energy storage and microgrids, work hard to foster integrated and optimized hybrid energy demonstration projects, develop technological innovations, infrastructures, operating mode and policy support system for the integrated development and utilization of new energy resources. China will work to develop the "Internet Plus" smart energy and build an energy Internet featuring coordinated development, integration and complementation of the energy "source-network-load-storage-user" with renewable energy as the main body.

1.5.1.4 Energy Technology Innovation Planning

According to *the 13th FYP Development Plan for Energy Technology Innovation*, we should promote the construction of energy Internet, strengthen the construction of smart power distribution and utilization networks, promote the utilization of distributed energy and hybrid power generation projects in the microgrids, and conduct research on the operation and trading of energy interconnection systems. We should construct energy Internet demonstration projects and promote multi-energy complementary and distributed hybrid power generation and microgrids.

1.5.2 Promotion of Hybrid Energy System Development from the Perspective of Compensation, Financing and Economic Support, etc.

1.5.2.1 Compensation (Market) Mechanism for Power Ancillary Services

(1) *Interim Measures for the Administration of Ancillary Services of Grid-connected Power Plants.*

The *Interim Measures for the Administration of Ancillary Services of Grid-connected Power Plants* (D.J.S.C. [2006] No.43) shall be applied to the power dispatching and trade centers at or above the provincial level as well as the grid-connected fire power and water power plants under the direct dispatch thereof. The document clearly defines and classifies ancillary services provided by grid-connected power plants. According to the document, the "paid ancillary services" refer to the ancillary services provided by grid-connected power plants other than basic ancillary services, and include

AGC, paid peak shaving, standby, paid reactive power regulation, and black-start services. The compensation shall be made for paid ancillary services. The document provides detailed provisions on the methods of measurement and assessment of paid auxiliary services, compensation ways and sources of expenses, and supervision and control measures.

(2) *Work Plan for Improving the Compensation (Market) Mechanism for Power Ancillary Services.*

In recent years, the electric power industry in China, especially clean energy, has witnessed rapid development. The power supply structure and grid structure have undergone major changes. The system scale has been continuously expanded. As a result, the complexity of system operation and management has greatly increased, putting higher demands on the safe and stable operation of the system. To optimize the distribution of resources in a wider area, there is an urgent need to further improve and deepen the compensation (market) mechanism for power auxiliary services. The *Work Plan for Improving the Compensation (Market) Mechanism for Power Ancillary Services* (G. N. F. J. G. [2017] No. 67) issued by National Energy Administration in November 2017 specified detailed stage development objectives and main tasks for ancillary services in the national power market and proposes to improve the existing related rules and regulations, establish the power ancillary services sharing mechanism for power users involved in the long-term power transactions and promote trans-provincial and trans-regional power auxiliary services compensation by classification. The document will promote the nationwide implementation of power auxiliary services in a scientific way and provide a mechanism guarantee for the hybrid energy development.

1.5.2.2 Financing and Economic Support

(1) *Implementation Opinions on Promoting Integrated and Optimized Hybrid Energy Demonstration Projects.*

According to the *Implementation Opinions of the National Development and Reform Commission and National Energy Administration on Promoting Integrated and Optimized Hybrid Energy Demonstration Projects* (F. G. N. Y. [2016] No. 1430), we should implement new pricing mechanism.

For terminal integrated energy supply demonstration projects, new pricing mechanisms such as electricity price, heat price and gas price in favor of increasing system efficiency will be implemented with market-oriented reform before the formation of energy price market mechanism. We should implement the scientific price system such as peak-valley price, seasonal price, interruptible price, high reliability price and two-part tariff system to promote and implement the pricing mechanism such as gas and electricity price linkage and guide the users of electricity and natural gas to take the initiative to participate in the demand-side management. The specific price policy and level shall be determined by

national and local competent price department according to their authority. As for wind-solar-hydro-thermal-storage hybrid energy demonstration projects, China coordinates two means, market-based price and modeling price, accelerate the construction of electric power and natural gas spot market and electric auxiliary service market, and improve the market mechanism of auxiliary services such as peaking shaving, frequency regulation and backup mechanisms. Before the formation of market-based price, China should implement pricing mechanisms for electricity, gas, and ancillary services that are beneficial to all types of power supply regulation.

The supporting policy shall be enhanced. The integrated and optimized hybrid energy demonstration projects certified by the state will preferentially use the total quota defined in the national energy plan for each province (region or city), such as thermal power capacity, renewable energy development scale and subsidies. The surplus power of wind-solar-hydro-thermal-storage hybrid energy demonstration projects after local consumption may preferentially participate in trans-provincial or trans-regional power transmission and consumption. Qualified integrated and optimized hybrid energy projects are the key investment targets in the energy field. For qualified projects, the additional subsidy of renewable electricity price could be applied according to relevant procedures. Each province (region or municipality) may provide relevant projects with specific supportive policies through initial investment subsidy or interest subsidy or opening special bond account and so on according to local realities.

Innovative management systems and business models should be created. We should actively support building integrated and optimized hybrid energy demonstration projects through PPP. In combination with electricity, oil and gas system reform, we should create innovative management and operation model for terminal integrated energy supply system and carry out the reform of electricity sales business. National Energy Administration should work with relevant departments to improve such technical standards and regulations as power (gas or heat) grid connection and grid-connected operation and coordinate the interests of all parties concerned including energy users, energy suppliers, power (gas or heat) grid operators, to solve problems such as grid connection for terminal integrated energy supply system and surplus power or heat connected to grids. Enterprises in charge of relevant electric power grids, gas grids and heat supply networks should provide convenient, in-time and accessible grid-connected and emergency reserve services and implement fair scheduling. The business model of innovative integrated energy supply system encourages the formation of integrated energy service companies by the shareholding or participation of power grids, gas, heat companies to engage in market-based energy supply and power sales, and actively implement market-based mechanisms such as contract energy management and comprehensive energy conservation services. We should accelerate the construction of an Internet-based smart energy information service platform to provide users with open, shared, flexible and intelligent integrated energy supply and value-added services.

(2) *Trial Measures for Promoting the Construction of Grid-connected Microgrids.*

According to the *Trial Measures for Promoting the Construction of Grid-connected Microgrids* (F. G. N. Y. [2017] No. 1339), after complementation, the new energy power generation projects within the microgrids should be included in the subsidy scope of renewable energy development fund based on relevant procedures and subject to state renewable energy power generation subsidy policies. Local governments are encouraged to provide supporting policies for microgrid development.

(3) *Guiding Opinions on Promoting Energy Storage Technology and Industry Development.*

According to the *Guiding Opinions on Promoting Energy Storage Technology and Industry Development* (F. G. N. Y. [2017] No. 1701), it is necessary to improve policies and regulations, form a pricing mechanism for energy storage application based on the study of electricity market construction, actively implement pilot projects for innovative application of energy storage, establish a compensation mechanism, promote energy storage to participate in pilot projects of compensation systems for electric auxiliary services, develop compensation regulatory mechanisms; guide social investment, support and guide more social capital to invest in energy storage industry by various financing modes; promote market reforms, establish flexible market-based trading mechanisms and pricing mechanisms for energy storage and others, and encourage energy storage to directly participate in market transactions.

1.5.3 Construction of Hybrid Energy Projects from the Perspective of Project Promoting

1.5.3.1 *Guiding Opinions on and Action Plan for Accelerating the Construction of Distribution Network*

According to the *Guiding Opinions on and Action Plan for Accelerating the Construction of Distributed Network* (F. G. N. Y. [2015] No. 1899, G. N. D. L. [2015] No. 290), it is necessary to conduct microgrids demonstration and application in areas with high requirements for urban power supply reliability and in remote areas such as rural areas and islands, and to build microgrids of such energy sources as PV, wind power and gas turbine.

1.5.3.2 *Guiding Opinions on Promoting the Construction of New Energy Microgrid Demonstration Project*

According to the *Guiding Opinions on Promoting the Construction of New Energy Microgrid Demonstration Project* (G. N. X. N. [2015] No. 265), new energy microgrid demonstration projects are carried out to explore the establishment of an integrated local electric power system for the power generation, transmission, distribution, storage and utilization of renewable energy with high volatility, explore new commercial operating models and new forms of business for electrical energy services and form a well-developed new energy microgrid technology system and management system. On-grid new energy microgrids are preferred in areas with high penetration rate of distributed renewable energy and multi-energy complementary conditions. Independent (or weakly interconnected) new energy microgrids are mainly used in remote areas and on islands not covered by grid and in areas only relying on small hydropower. It may also be a transformation of a village independent solar power plant that has been completed but has insufficient power supply capacity in areas with the project of power transmission to townships or areas without electricity.

1.5.3.3 *Guiding Opinions on Promoting the "Internet Plus" Smart Energy Development*

According to the *Guiding Opinions on Promoting the "Internet Plus" Smart Energy Development* (F. G. N. Y. [2016] No. 392), it's necessary to promote the coordinated development of centralized and distributed energy storage, and develop multi-type, large-capacity, low-cost, high-efficiency, long-life energy storage products and systems such as electricity storage, heat storage, cooling storage and clean fuel storage; to help centralized new energy power generation base be equipped with appropriate-scale energy storage power plants to achieve coordinated and optimized operation of energy storage systems and new energy sources or power grids. We must strengthen the building of an integrated hybrid energy network and build integrated energy network based on smart grids, connecting with various types of networks such as heat supply networks, natural gas pipelines and transportation networks, and featuring the coordinate conversion of multiple energy sources and coordinated operation of centralized and distributed energy sources. It's necessary to promote the building of infrastructures supporting the flexible conversion, efficient storage and intelligent coordination of electricity, cooling, heating, gas, hydrogen and others.

1.5.3.4 *Implementation Opinions on Promoting Integrated and Optimized Hybrid Energy Demonstration Projects*

According to the *Implementation Opinions of National Energy Administration on Promoting Integrated and Optimized Hybrid Energy Demonstration Projects* (F. G. N. Y. [2016] No. 1430), it is one of the important tasks in building an "Internet Plus" smart energy system to construct integrated and optimized hybrid energy demonstration projects. The main task is to build a terminal integrated energy supply system and a wind-solar-hydro-thermal-storage hybrid energy system. It clarifies the main tasks, construction goals, construction principles and methods, policy measures, and implementation mechanisms for integrated and optimized hybrid energy demonstration project construction. Policy measures include implementing new pricing mechanism, increasing policy support and making innovative management system and business models.

1.5.3.5 *Notice on Announcing the First Batch of Integrated and Optimized Hybrid Energy Demonstration Projects*

According to the *Notice of National Energy Administration on Announcing the First Batch of Integrated and Optimized Hybrid Energy Demonstration Projects* (G. N. G. H. [2017] No. 37), National Energy Administration organized the review and certification of integrated and optimized hybrid energy demonstration projects and identified the first 23 demonstration projects (see Table 1.5.1). The first batch of demonstration projects should start in principle before the end of June 2017 and be completed and put into production by the end of 2018. Projects that have been included in the first batch of demonstration projects will be canceled if they do not meet the demonstration requirements due to major changes, or if they fail to start before the end of June 2017.

There are 17 demonstration projects for terminal integrated energy supply systems, including one in Beijing, one in Shanxi, one in Anhui, one in Inner Mongolia, one in Shandong, one in Guangdong, one in Hubei, and one in Xinjiang, two in Jiangsu, three in Hebei, and four in Shaanxi. Currently, demonstration projects are mainly carried out in new urban construction areas, economic development zones, industrial parks, etc. The overall development and mix use of traditional energy sources and new energy sources should be implemented according to local conditions. The energy smart microgrids are built and heating, electricity, water and cooling are supplied in a coordinative way, to realize the integration of energy supply and consumer terminals.

There are six demonstration projects of wind-solar-hydro-thermal-storage hybrid energy system, including one in Hebei, one in Inner Mongolia, one in Sichuan, one in Shaanxi and two in Qinghai. In combination of local resources, demonstration projects combine multiple energy resources, including wind power, PV, solar thermal energy, energy storage, hydropower, thermal power and others, to realize efficient and complementary utilization, improve the coordinate ability of energy supply and demand and promote clean energy production and consumption.

1.5.3.6 *Notice on Issuing the List of New Energy Microgrid Demonstration Projects*

According to the *Notice of National Energy Administration on Issuing the List of New Energy Microgrid Demonstration Projects* (F. G. N. Y. [2017] No. 870), National Development and Reform Commission organized an expert review to determine 28 new energy microgrid demonstration projects. The demonstration projects focused on the integrated application of technologies as well as innovative operation management models and market-based transaction mechanism. List of New Energy Microgrid Demonstration Projects is shown in Table 1.5.1.

Table 1. 5. 1 List of New Energy Microgrid Demonstration Projects

S. N.	Project Name	Geographical location	Scope of Energy Supply
Grid-connected			
1	Yanqing New Energy Microgrid Demonstration Project in Beijing	Beijing City	1. South Area of Badaling Economic Development Zone; 2. Beijing Humanities University and its surrounding areas; 3. Heating Center of Badaling Economic Development Zone and its surrounding area; 4. Kangzhuang Town Industrial Development Zone; 5. Badaling Scenic Area; 6. Solar Thermal Experimental Power plant of Chinese Academy of Sciences
2	New Energy Demonstration Park of Xishan Ecological Industrial District in Taiyuan	Xishan Ecological Industrial District in Taiyuan City, Shanxi Province	20 parks in Xishan Ecological Industrial District in Taiyuan City, Shanxi Province
3	New Energy Microgrid Demonstration Project of Zhangbei Cloud Computing Base Green Data Center	Zhangbei Cloud Computing Base in Hebei Province	Zhangbei Cloud Computing Base Area
4	Microgrid Demonstration Project of Hefei High-tech Development Zone	Hefei National High-tech Development Zone in Anhui Province	Hefei National High-tech Development Zone
5	New Energy Microgrid Demonstration Project of Baicheng Industrial Park in Jilin Province	Baicheng Industrial Park in Jilin Province	22 km^2 Baicheng Industrial Park
6	Wind-solar-hydrogen-storage Smart Microgrid	Shaanxi Province	Industrial Park transformed from the old plant of Shaanxi Baoguang Group, covering an area of 100 mu
7	Smart Microgrid Demonstration Project of Macaoenergy Industrial Park	Industrial Park in Bijie City, Guizhou Province	Jinhaihu New District of Bijie Experimental Area in Guizhou, covering an area of 500 mu
8	New Energy Microgrid Demonstration Project of Haidian Northern New Area in Beijing	Beijing City	Cuihu and Yongfeng Area in the North of Haidian, with the power supply area of 879,000 m^2
9	Key Technical Research and Demonstration and Application Project of SGCC Jiaxing New Energy Microgrid	Jiaxing City in Zhejiang Province	Jiaxing Qinfeng Substation

Continued

S. N. Project Name	Geographical location	Scope of Energy Supply	
10	Ubiquitous Energy Internet of Sino-German Ecopark Initiating Zone	Qingdao City in Shandong Province	Sino-German Ecopark, covering an area of about 100,000 m^2
11	New Energy Microgrid of Jinan iESLab Industrial Park in Shandong	Jinan Jicheng Industrial Park in Shandong Province	iESLab Industrial Park, covering an area of 308 mu
12	Smart Microgrid Demonstration Project of Lingang Campus of Shanghai University of Electric Power	Shanghai City	Lingang Campus of Shanghai University of Electric Power, covering an area of about 960 mu and with the floor area of 575,000 m^2
13	Qingdao Dongjiakou Port New Energy Microgrid Demonstration Project	Qingdao City in Shandong Province	Power supply for Lingang Industrial Park; main industrial users in the park involve those engaged in fossil, metallurgy, equipment manufacturing
14	New Energy Microgrid Demonstration Project of Tai'an Taikai Southern Area Industrial Park	Tai'an Taikai Southern Area Industrial Park in Jiangsu	Mainly supply power for the office building, guard room, factory supervision and control system and No. 9 Manufacturing shop of Taikai Packaged Substation Co., Ltd. in Taikai Southern Area Industrial Park
15	Tianchang Meihao Village Smart Microgrid	Tianchang City in Anhui Province	Erdun Village, Yongfeng Town, Tianchang City
16	Jiaze Hongsibu New Energy Smart Microgrid Project in Ningxia	Hongsibu District, Wuzhong City, Ningxia	Hongde Industrial Park in Hongsibu District, Wuzhong City, Ningxia
17	Kelu Smart Grid Experimental Demonstration Project	Yumen Economic Development Zone in Gansu Province	Yumen Economic Development Zone
18	Chongli Olympic Zone New Energy Microgrid	Zhangjiakou City in Hebei Province	Yunding Ski Park + Zhangjiakou Mountain Media Center, Zhangjiakou Olympic Village, Nordic Center and Biathlon Center as well as Taiwu Four-season Tourist Resort, with an area of 854,500 m^2
19	Low-carbon City Oriented Chongli Group Microgrid Demonstration Project	Zhangjiakou City in Hebei Province	Chongli City Central City Area, Olympic Zone and Towns Surrounding Chongli

Continued

S. N. Project Name	Geographical location	Scope of Energy Supply	
20	Microgrid Demonstration Project of Wenzhou Economic and Technological Development Zone	Wenzhou Economic and Technological Development Zone in Zhejiang Province	Binhai New District and Jinhai Garden of Wenzhou Economic and Technological Development Zone, covering an area of 4 km^2
21	New Energy Microgrid Demonstration Project of Suzhou Xiexin Industrial Application Institute	Suzhou City, Jiangsu Province	Suzhou Xiexin Industrial Park
22	New Energy Microgrid Demonstration Project of South Garden of Jinan Economic Development Zone	Jinan Economic Development Zone in Shandong Province	Shandong Hailun Environmental Protection and Technology Co., Ltd., Shandong Haonuo Medical Co., Ltd., Shandong Bolin Environmental Protection and Technology Development Co., Ltd.
23	Jiuquan Suzhou District New Energy Microgrid Demonstration Project in Gansu Province	Suzhou District, Jiuquan City, Gansu Province	Phase I of Suzhou District Comprehensive Utilization and Development Zone, covering an area of 5.5 km^2
24	Nansha High reliable, smart and low-carbon microgrid demonstration project of Guangzhou Administration of Power Supply	Nansha Training Base of Guangzhou Administration of Power Supply in Guangdong Province	Supply power to Nansha Training Base, covering an area of 3,000 m^2
		Independent	
1	Zhaizhushan Island New Energy Microgrid Project in Zhoushan	Zhaizhushan Island in Zhoushan	Supply power to inhabitants and public utilities on the island
2	Beilong Island Solar-storage-diesel hybrid power microgrid demonstration project in Ruian City	Ruian City in Zhejiang Province Beilong Island	Supply power to inhabitants and public utilities on the island
3	Taishan Island Wind-solar-diesel-storage Integrated Project in Fuding	Fuding City in Fujian Province Taishan Island	Supply power to inhabitants and public utilities on the island
4	Smart Microgrid Demonstration Project of Wanshan Island in Zhuhai	Zhuhai City in Guangdong Province Wanshan Island	Supply power to inhabitants and public utilities on the island

Chapter 2

Potentials of Development of Hybrid Energy Systems in ASEAN Region

2.1 Geographic Overview

Located in the southeastern part of Asia, the ASEAN is at the "crossroad" between Asia and Oceania and the conjunction between the Pacific Ocean and the Indian Ocean. The ASEAN members include Brunei, Cambodia, Indonesia, Laos, Malaysia, Myanmar, the Philippines, Singapore, Thailand and Vietnam. The total land area is about 4.48 million km^2 and the length of continental coastline is more than 10,000 km.

The ASEAN nations have numerous islands. Indonesia is the largest archipelagic nation in the world, with 17,508 islands; the Philippines has over 7,100 islands; Singapore consists of Singapore Island and the neighboring 63 islands; Brunei has 33 islands.

2.2 Energy Resources

Fossil energy is the most primary energy resource in the ASEAN, including natural gas, petroleum, hard coal and lignite, which are mainly distributed in Indonesia, Malaysia, Vietnam and Thailand. The economic exploitable volumes of the aforesaid four types of resources are 6.46 trillion m^3, 1.82 billion tons, 37.53 billion tons and 10.23 billion tons that respectively account for 39%, 29%, 38% and 3% of the gross reserves. Refer to Table 2.2.1 for the overview of fossil energy in the ASEAN.

Table 2.2.1 Overview of Fossil Energy Resources in the ASEAN

Item	Natural gas /(10^{12} m^3)	Crude oil /(10^8 t)	Hard coal /(10^8 t)	Lignite /(10^8 t)
Gross reserve	16.74	63	983.4	3,534.3
Exploitable reserve	6.46	18.2	375.3	102.3

The ASEAN boasts abundant resources of renewable energy especially the hydro energy, biomass energy and geothermal energy with potentials above 240 GW, 37 GW and 33 GW respectively. The solar energy, wind energy and tidal energy are also promising. The average daily irradiation of solar energy is around 5 kWh/m^2 and the potentials of wind energy and tidal energy are above 87 GW and 219 GW, respectively. However, the resource endowment and development conditions among the ASEAN nations vary a lot. Indonesia has the most abundant diversity of renewable energy and the richest resource reserves especially the hydroenergy resource; Myanmar, Vietnam, Malaysia and Laos are mainly advantageous in terms of hydroenergy resource; the Philippines has more resources of tidal energy and wind energy; Thailand enjoys abundant wind energy resource. Refer to Table 2.2.2 for the overview of renewable energy resources in the ASEAN.

Table 2.2.2 Overview of Renewable Energy Resources in the ASEAN

Country	Biomass energy /GW	Geothermal energy /GW	Hydro energy /GW	Wind energy /GW	Tidal energy /GW	Solar energy /[kWh/(m^2 · d)]
Brunei			0.07			9.6 – 12
Cambodia			10			5
Indonesia	32.6	28.9	75		49	4.8
Laos	1.2	0.05	26			3.6 – 5.3
Malaysia	0.6		29			4.5
Myanmar			40.4	4		5
the Philippines	0.24	4	10.5	76	170	5
Singapore					0.03 – 0.07	3.15
Thailand	2.5		15			5 – 5.6
Vietnam	0.56	0.34	35	7	0.1 – 0.2	4.5

2.3 Status Quo and Demand of Electric Power

In 2016, the ASEAN's total energy consumption reached 634 million tons of standard coal where the consumption of clean energy accounted for about 16%. The total installed capacity of ASEAN was 217,000 MW, among which renewable energy was 53,000 MW, accounting for 24.6%. The total power output reached 962.5 TWh including 196.8 TWh of renewable energy power output, a proportion of 20.5%. Refer to Figures 2.3.1 and 2.3.2 for the sketches of ASEAN's energy consumption structure and installed power capacity.

On the basis of the economic development targets and actual development situation of

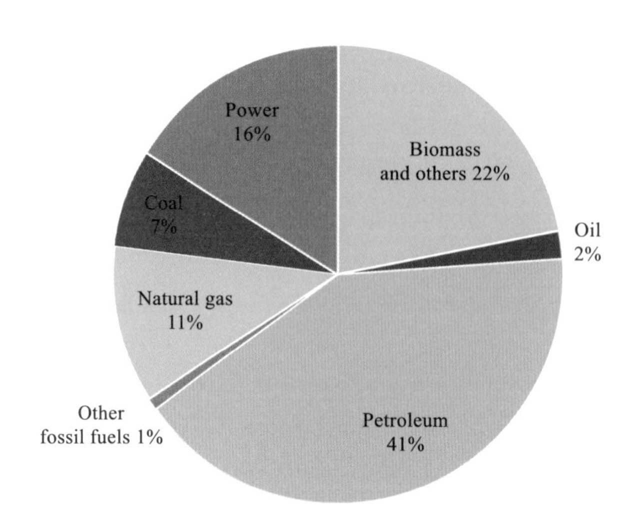

Figure 2. 3. 1 Sketch of ASEAN's Energy Consumption Structure

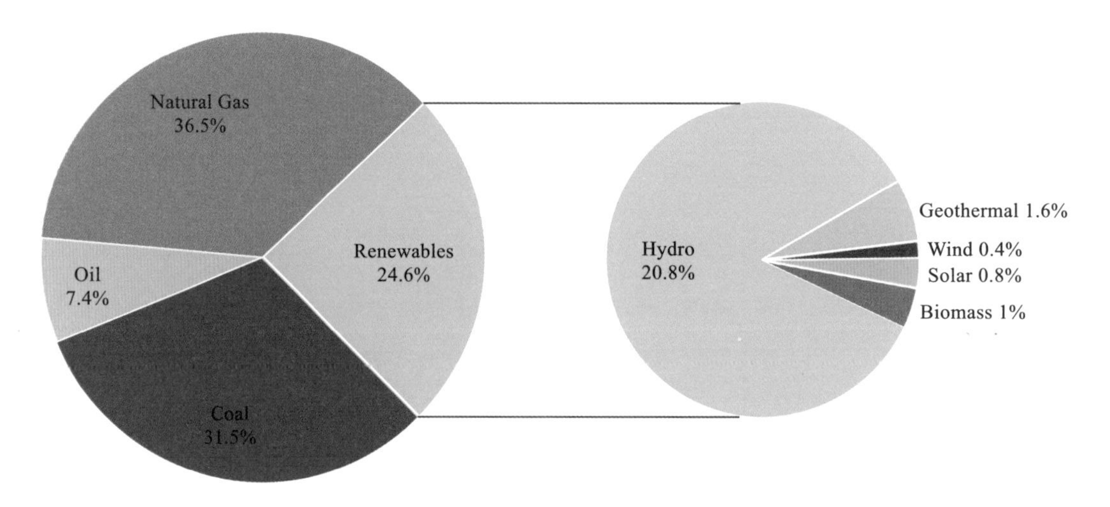

Figure 2. 3. 2 Sketch of ASEAN's Installed Power Capacity

ASEAN member nations, the electric demand forecast has been conducted in accordance with different growth paces of economy. Refer to Figure 2. 3. 3 for the results of forecast. In accordance with the program of medium-speed economy growth (5. 3% annually), in 2020, the ASEAN's electric power demand will amount to 1,003. 9 TWh with an annual growth rate of 4. 6%. Refer to Figure 2. 3. 4 for the forecast results of electric power demand under the medium growth speed program.

The analysis of supply and demand balance of the ASEAN nations in 2020 has been made in line with the forecast results of electric power demand under the medium economic growth speed program. The analysis results show that among the nations with higher power demand, Indonesia will have surplus power and thus is capable of power export; Thailand and the Philippines will face more or less power shortage and thus need the support of

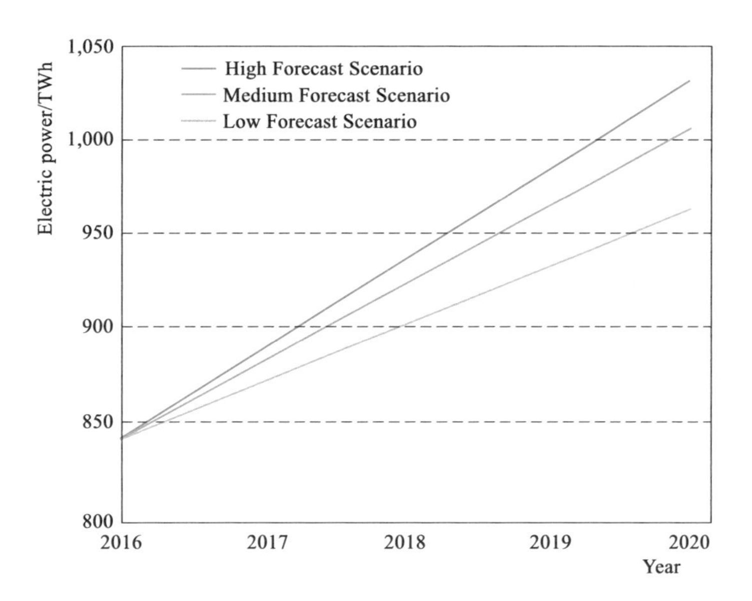

Figure 2. 3. 3 Forecast of ASEAN's Electric Power Demand
(Scenario of Different Economic Growth Speed)

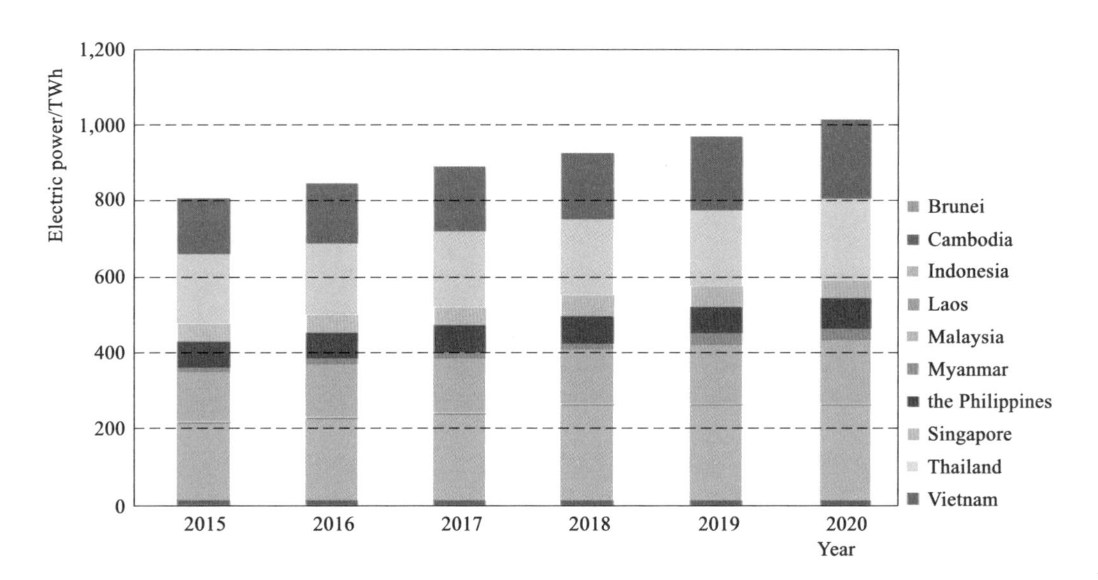

Figure 2. 3. 4 Forecast of ASEAN's Electric Power Demand
(Scenario of Medium-speed Economic Growth)

neighboring countries; Vietnam and Malaysia will basically reach balance between supply and demand. Among the nations with lower power demand, Laos and Myanmar may supply power to the neighboring countries due to their abundant hydroenergy resource; Cambodia will make both ends meet; Singapore and Brunei may consider importing power generated from renewable energy.

2.4 Development of Renewable Energy

To achieve the target clarified at the 33rd ASEAN Ministers on Energy Meeting in 2015, i.e. the renewable energy should account for 23% of gross primary energy consumption in the ASEAN, all ASEAN Member States (AMS) have set their own targets of renewable energy development. Refer to Table 2.4.1.

Table 2.4.1 Targets of Renewable Energy Development of ASEAN Countries

Country	Target of renewable energy power development
Brunei	Till 2035, the proportion of total installed capacity of renewable energy will reach 10%
Cambodia	
Indonesia	Till 2025, the renewable energy supply will account for 23% of total primary energy supply
Laos	Till 2025, the renewable energy will account for 30% of total end energy consumption
Malaysia	Till 2020, the installed capacity of renewable energy will reach 2,080 MW (exclusive of the hydropower above 30 MW)
Myanmar	Till 2030, the installed capacity of renewable energy power will reach 1,100 MW, accounting for 47% of total installed capacity
the Philippines	Till 2030, the installed capacity of renewable energy power will be twice of that in 2010, reaching 15,300 MW
Singapore	Till 2020, the solar power will reach 350 MW. Till 2018, the waste energy recovery will be 10,140 t/d
Thailand	The renewable energy will account for 30% of total energy consumption and 15%-20% of power generation
Vietnam	Till 2020, the installed capacity of renewable energy power will account for 11% of total installed capacity; such proportion will reach 13% in 2025 and 21% (exclusive of the hydropower above 30 MW) in 2030 respectively

According to the 5th *ASEAN Energy Outlook (2015 - 2040)*, ACE made three predictions for the development of renewable energy in the ASEAN by 2040, including Business as Usual (BAU), Analysis of Target Situation (ATS, a comprehensive analysis based on the emission reduction and renewable energy development objectives of the ASEAN), Active Protection System (APS, i.e. active energy efficiency and renewable energy policy). The results of prediction are shown in Figure 2.4.1.

According to the results, the ASEAN countries will embrace rapid development of renewable energy. Under the APS, the installed capacity of hydropower, PV and wind power will reach 144 GW, 102.7 GW and 27.3 GW respectively by 2040, registering 3 times, 59 times and 31 times those in 2016 respectively.

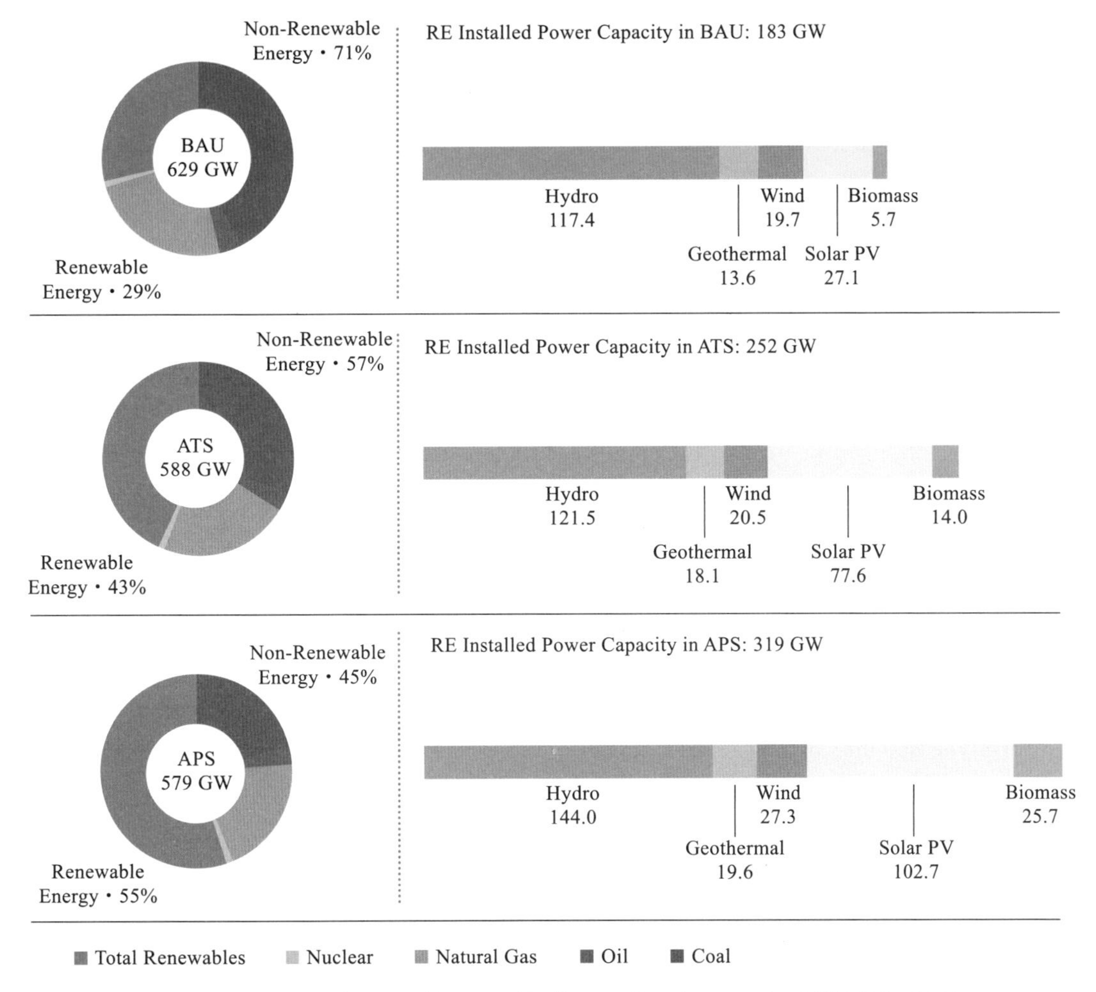

Figure 2. 4. 1 Prediction of Renewable Energy Development in ASEAN by 2040

Source: the 5th *ASEAN Energy Outlook* (*2015 - 2040*), ACE

2. 5 Prospect of Hybrid Energy System Application

In April 2016, the leaders of more than 170 nations jointly signed the *Paris Agreement*, committed to control the global temperature rise within 2℃. We are on a new starting point of global climate control. Facing the new energy situation, the international society has a common mission of developing renewable energy. In accordance with the *BP Statistical Review of World Energy* released in June 2017, the growth rate of global renewable energy (including biomass fuel) in 2016 hit 12%, the fastest among all energy sources. Geographically, the Asia-Pacific region turned out to be the world largest producer of renewable energy.

The ASEAN nations have rich fossil energy so that the present electric power structure relies on thermal power. But the burning of fossil energy in a long run will lead to a series of environmental problems like air pollution and greenhouse gas effect, lowering the living standards of local residents. To satisfy the swelling power demand in the ASEAN nations and achieve good interaction between economic development and ecological environment, it is necessary to uplift the proportion of clean energy and transform the energy structure.

2.5.1 The Existing Problems of Energy Development and the Estimated Future Tendency in the ASEAN Nations

In 2016, there were 1.1 billion people without access to electric power in the world and that number in Asia was 455 million. Therefore, it is of extreme importance to enhance energy accessibility. As of 2016, Thailand and Vietnam had basically realized energy accessibility and the Philippines, Indonesia and Laos had completed 90%. However, Cambodia and Myanmar only registered 60% energy accessibility. It is of urgent need to extend the electric power accessibility and mitigate energy poverty.

To satisfy electric power demand and respond to the renewable energy proportion target set at the 33rd ASEAN Ministers on Energy Meeting, the ASEAN nations are welcoming a booming period of wind and solar power. Given the characteristics of wind and solar power output, i.e. intermittence, volatility and randomness, large-scale grid connection of wind and solar power may harm the stability of power grid, and, if the peak shaving capacity is inadequate, such wind and solar power may be more or less curtailed.

Indonesia has long and narrow territories with the south-north span of merely 1,888 km while the east-west span as long as 5,110 km. Given the peculiar geographic structure of an island-based nation, its electric power system consists of Java-Bali power grid, Sumatra power grid and east Indonesian power grid with the possibility of interconnection of these grids in future. Additionally, as the energy demand of ASEAN nations is growing rapidly, the imbalance between supply and demand and the intent of extensive cross-border collaboration means the opportunities of electric power cooperation. At present several transnational power transmission lines have been built among several ASEAN nations and such power connectivity will be further enhanced.

2.5.2 The Prospects of Hybrid Energy System Application in the ASEAN Nations

2.5.2.1 Development of Microgrids for Wider Coverage of Electric Power

The ASEAN nations have numerous islands. For the remote and small islands that are less populated, the electric power coverage by means of undersea cables would be too costly or even economically unaffordable. The development of microgrids in line with local conditions can effectively save investment and expand power coverage, which will be quite promising. For instance, the microgrid project at Dongfushan Island in China adopts the hybrid energy supply pattern by using local abundant renewable clean energy as the primary power source, which is complemented by diesel power generation and its

fresh water supplied by seawater desalination system, having solved the problem of water and power supply to the island inhabitants. The project had been put into operation for 7 years with satisfactory overall performance.

2.5.2.2 Mutual Complementation between New Energy and Regulating Power Source for a Higher Efficiency of Energy Utilization

With the accelerated and scale-up development of wind and solar power, the wind-solar power and regulating power source can suppress the output volatility of new energy, protect the safety and stability of power grid and improve the energy utilization efficiency and economy. On the basis of electric power development need of the ASEAN nations and in line with the characteristics of rich resources of natural gas, oil, coal and renewable energy, and driven by the philosophy of promoting renewable energy growth, the hybrid energy systems such as hydro-solar hybrid, hydro-wind-solar hybrid, wind-solar-coal hybrid, pumped storage energy for new energy operation, etc. will have considerable potential of development. For instance, Longyangxia hydro-solar hybrid power plant in China boasts a total installed capacity of 850 MWp. With the hydropower and solar power hybrid system, Phase Ⅰ 320 MWp has witnessed an increase of 10% of annual utilization hours for the transmission line; after overall grid connection of 850 MWp, the annual utilization hours will increase by 22.4%. The hydro and solar hybrid energy system has therefore enhanced the efficiency of output transmission lines.

2.5.2.3 Bundled Transmission of Multiple Energy to Solve the Problem of Long-distance Energy Allocation

Some ASEAN countries will require long-distance energy allocation, which offers a great space for developing multi-energy bundled transmission so as to improve the energy transmission efficiency and make better use of power transmission channels. In addition, increasing inter connection of electric power grids between the ASEAN nations also creates favorable conditions for the bundled transmission of multiple energy resources such as wind-solar-thermal energy and wind-solar-pumped storage, etc. For instance, the Hami-Zhengzhou ±800 kV UHV DC wind-solar-thermal power bundled transmission project in China with a power transmission capacity of 8,000 MW performs bundled transmission of thermal power, wind power and solar power of Hami, Xinjiang Autonomous Region. It is featured by long distance, heavy duty, little loss, environment friendliness and land saving, contributing to the development and utilization of large-scale energy bases and mitigating the power shortage in the central and eastern parts of China.

Chapter 3

Application of Hybrid Energy System

3.1 Functions of Different Kinds of Power Sources in Hybrid Energy System

The hydropower units are featured by flexible start/stop, quick response and unlimited times of start/stop and are thus used for peaking shaving, frequency and phase regulating and as standby power source. Their peak shaving capacity is up to their installed capacity. Making use of the regulating and quick response capabilities of reservoirs, the hydropower plants can compensate and regulate new energy, suppress the excessive volatility of new energy and satisfy the demand of power consumption. By storing and releasing water through reservoirs, the hydropower plants with peak shaving capability can adjust its power output to complement the operation of new energy power plants.

The wind and solar power plants provide clean electric power but the available capacity of wind and solar power is close to zero with frequent change of output, which is extremely volatile, intermittent and uncontrollable. In actual operation, most wind power has anti-peak-shaving feature that requires compensation and regulation by other capable power plants. The solar-thermal power generation may generate a smooth output curve by two means: heat storage and natural gas compensation. Its operation performance is similar to a fossil-fueled thermal power. It has almost no impact on the safety and stability of power grid and can be easily consumed by the grid.

The thermal power is a regular flexible power source that offers both capacity and power to the system with slow change in output (less flexible than hydropower). The thermal power unit generates power in a stable way but with longer start/stop duration. When load changes, only by burning extra coal can the output of thermal power unit climb up. Once economically permissible, the thermal power units will be helpful to solve the problem of new energy consumption to some extent by properly uplifting the peak shaving rate. When acting as the independent regulating power source to assist in new energy operation, as the thermal power reacts more slowly than hydropower, it can hardly accommodate the frequency change of wind (solar) power and thus requires a wind (solar) power forecast system and other auxiliary power sources.

The pumped storage power plant is a reliable peak shaving and standby power source that has quick response to load change and flexible operation regulating. It may be used for diverse purposes like peak shaving, valley filling, frequency regulating, phase regulating and emergency reserve. Its peak shaving capacity is twice as much as the workload. This serves as an important tool to guarantee the grid safety and stability. Thanks to its energy storage function, the pumped storage power plant can store and convert the wind or solar energy and make use of extra electric power to pump water, which is highly complementary to the new energy. It helps make the wind (solar) power output less volatile, the power transmission system more flexible and the new energy consumption capacity larger at the receiving end of power grid.

The energy storage technology is an effective means to improve the comprehensive energy utilization efficiency. Depending on the different forms of electric power conversion storage, the energy storage technology can be categorized into four types namely physical storage, chemical storage, electromagnetic storage and phase change storage. The energy storage effect is mainly demonstrated as less intermittent power fluctuation, higher utilization efficiency of new energy and additional reserve capacity as well as improved safety, stability and power supply quality of grids.

3.2 Conditions for Hybrid Energy System Application

3.2.1 Conditions for Hydropower and New Energy Hybrid System Application

(1) The hydropower plant has the regulating capacity for a day or above (regulating the water inflow in a day by reservoir and discharging for power generation as required) with no other comprehensive utilization tasks like flood prevention, irrigation and water supply.

(2) It has new energy resource in the vicinity or in the power grid that it serves.

(3) It has market space for electric power especially in the location lack of electricity.

(4) It has a perfect grid connection system.

Case: With an installed capacity of 1,280 MW and reservoir regulating capacity of 19.35 billion m^3, China's Longyangxia Hydropower plant has multi-year regulating capability and perfect grid connection system totaling 6 feed-out bays including 5 feed-out bays in service while 1 feed-out bay standby. The solar power plant is located in the Solar Power Park of Gonghe County, Qinghai Province and its direct distance from Longyangxia Hydropower plant is merely around 30 km. With a total installed capacity of 850 MWp, this solar power plant is connected with the spare bay of Longyangxia Hydropower plant and gets compensated by reservoir regulating. The combined output of these two power sources is fed to the power grid via the transmission channel of Longyangxia Hydropower plant. The Longyangxia hydro-solar hybrid power project mainly supplies power to the grid of Qinghai Province and has the corresponding power output market space.

3.2.2 Conditions for Thermal Power and New Energy Hybrid System Application

(1) The resources for thermal power such as coal and natural gas are available and abundant.

(2) A thermal power plant with strong peak shaving and frequency regulating capability is built or planned, which does not concurrently supply heat (which affects the peak shaving capability).

(3) There are new energy resources in the vicinity or in the power grid that it serves.

(4) There is a market space for electric power especially in the region lack of electricity.

(5) There is a perfect grid connection system.

Case: Xinjiang Zhundong region has rich coal, wind and solar energy resources and the thermal power is one of the backbone industries for local economy development. To improve new energy consumption and mitigate the uneven distribution of domestic resource and power consumption load, the Xinjiang Zhundong-Anhui UHVDC power transmission project was approved in form of bundled wind-solar-thermal power. Huge amount of wind (solar) power planned in Xinjiang Zhundong region will be delivered to higher consumption—Anhui Province of China. As the ancillary power source to the transmission project, its thermal project helps suppress the fluctuation of wind and solar power output, protect the frequency stability and safety of electric power system and promote new energy consumption.

3.2.3 Conditions for Pumped Storage Energy and New Energy Hybrid System Application

(1) There are natural resource conditions for construction of a pumped storage power plant.

(2) Other regular power sources in the power grid like hydropower and thermal power have given full play to their own capacity.

(3) There are new energy resources in the vicinity or in the power grid that it serves.

(4) There is market space for electric power especially in the region lack of electricity.

(5) There is a perfect grid connection system.

(6) There is no scale-up development condition for other energy storage facilities.

(7) The pricing mechanism is good (pricing mechanism of capacity price, frequency and phase regulating, standby, energy storage, etc.).

Case: Gansu Province has rich resources of wind and solar energy. The Chinese Government approved the construction of clean energy bases in two cities, such as Jiuquan and Jiayuguan on the Hexi Corridor, together with Jiuquan-Hunan ±800 kV UHVDC transmission project to solve the local new energy consumption problem. Given the gigantic scale of new energy construction on the Hexi Corridor, the energy storage facility of excellent peak shaving performance—pumped storage power plant can suppress the new energy output fluctuation, sustain the safe and stable operation of transmission system and make the long-distance power transmission more economic.

3.2.4 Conditions for Energy Storage and New Energy Hybrid System Application

(1) There are new energy resources in the area.

(2) There is market for electric power especially in the region lack of electricity.

(3) There are the conditions for the development of energy storage facility.

Case: Shuanghu County is situated at northern Tibet Qiangtang Plateau of northwest Naqu Prefecture of Tibet Autonomous Region. Before 2015, Shuanghu county seat solely depended on the power supply by a 220 kW solar power plant and one 450 kV diesel generator with limited installed capacity and output. After full consideration of local abundant solar energy resource and the environmental and meteorological conditions, a local independent renewable energy grid system was established primarily dependent on solar power with energy storage technology, which realizes the coordinated control and operation of power sources, grid and load under an exclusive electric power system.

3.2.5 Conditions for Pure New Energy Hybrid System

3.2.5.1 Hybrid System for Pure Wind Power Plants and Hybrid System for Pure Solar Power Plants

(1) The region has rich new energy resources and huge development potential.

(2) The region has centralized resource scale.

(3) The region is far away from the load center.

(4) The region can hardly afford the high construction cost of transmission lines, but with urgent need of development and utilization of new energy resources.

Case: Hami of Xinjiang Autonomous Region and Jiuquan of Gansu Province both have rich resources of wind and solar energy. Based on their resource endowment, they build large scale wind and PV power bases in the desertified land and through the mix of pure wind power and the mix of pure solar power to reduce the volatility of wind and solar power output.

3.2.5.2 Hybrid System for Wind Power and Solar Power

(1) The region has rich wind and solar power resources.

(2) To facilitate construction and operation management, the wind and solar energy resources are developed by the same investor.

(3) The region can hardly afford high construction cost of transmission channels and has rare corridors but urgent need of development and utilization of new energy resources.

(4) The region is short of land resources.

Case: During the construction of Xinjiang Hami wind power base, certain scale of solar power plant has been arranged in the inter-wind zone that not only saves land resource but also improves the utilization rate of power transmission lines by wind and solar shared transmission lines.

3.3 Modes of Hybrid Energy Systems

3.3.1 Hybrid Systems at Power Source Side

3.3.1.1 Hybrid System for "Hydropower-Wind (Solar) Power"

(1) Characteristics.

1) The new energy is deemed as a "virtual hydropower unit" that is connected with the hydropower plant and after being regulated by turbine generator units, the combined output of two power sources is fed to the power grid via the transmission channel of the hydropower plant.

2) During the idle time of power consumption of the electric power system, in case of larger output of new energy, the hydropower plant will properly reduce its output and store water in its reservoir. During the peak time of power consumption, the reservoir discharges water and the hydropower plant uplifts its power output. In other words, the new energy power is converted to reservoir storage and then temporally reallocated.

3) The quick response of hydropower can smooth the frequency change of new energy output and reduce the impact of new energy upon the power grid.

(2) Functions.

1) It mitigates the poor stability and output volatility of power generation by new energy from the power source side, protecting the safety and stability of power grid.

2) It makes the unstable and intermittent new energy output more acceptable to the power grid, and turns the low-quality electric energy into the high-quality electric energy that yields higher economic return.

3) It saves the investment in power transmission system or improves the utilization rate of transmission lines and new energy as well.

4) It promotes the scale-up development of new energy and satisfies the power consumption demand.

5) It optimizes the resource allocation and propels long-distance and interregional consumption of new energy.

Refer to Figure 3.3.1 for the sketch of daily operation mode of hydropower and new energy without and with a hybrid system.

(3) Construction model.

1) On the basis of the existing invested hydropower, a solar power or wind power plant will be built in the vicinity and connected with the same power transmission system.

2) On the basis of the existing hydropower in the grid, a new energy base will be

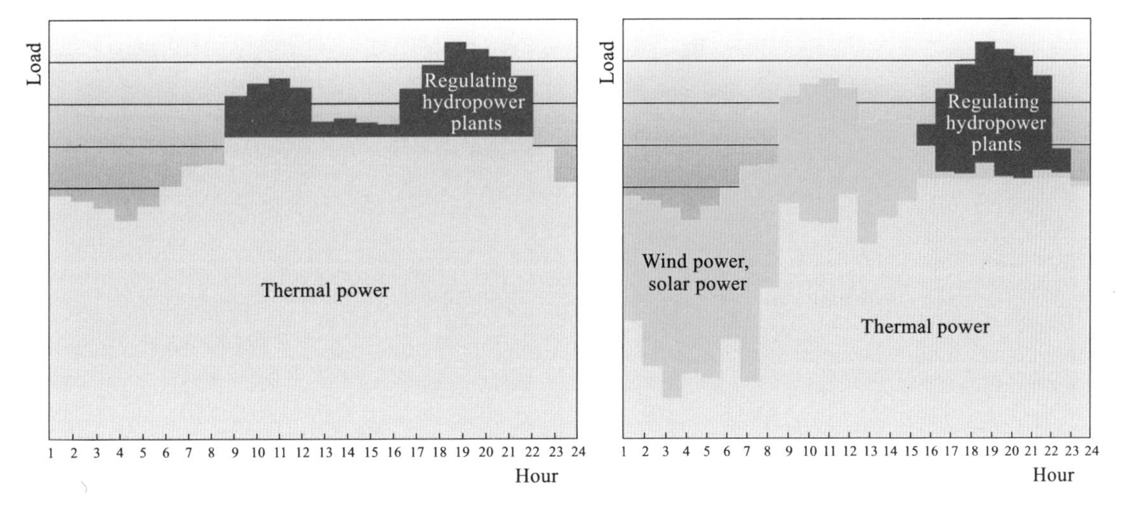

Figure 3. 3. 1 Sketch of Daily Operation Mode with or with out Hydropower and New Energy Hybrid System

built in the area with abundant resources and favorable development conditions and be connected with the same power grid.

3) The hydropower and comprehensive clean energy base will be planned as a whole. The power output is bundled and transmitted via the UHV channel to the load centers or power consumption areas.

3. 3. 1. 2 Hybrid System for "Thermal Power-Wind (Solar) Power"

(1) Characteristics.

1) When the electric power system is in idle time of consumption but with larger new energy output, the thermal power plant properly reduces its output. In the peak time of power consumption or when the new energy output shrinks, the thermal power plant increases its output.

2) The thermal power plant makes use of its peak shaving and frequency regulating capability to mitigate the impact of new energy output on the power grid, which increases new energy consumption and decreases fossil energy consumption.

(2) Functions.

1) It reduces output volatility and protects the safety and stability of power grid.

2) It saves investment in the power transmission system or improves the utilization rate of transmission lines and new energy as well.

3) It promotes the scale-up development of new energy and satisfies the power consumption demand.

4) It spurs long-distance and interregional consumption of new energy.

Refer to Figure 3. 3. 2 for the sketch of daily operation mode of thermal power and new energy without and with a hybrid system.

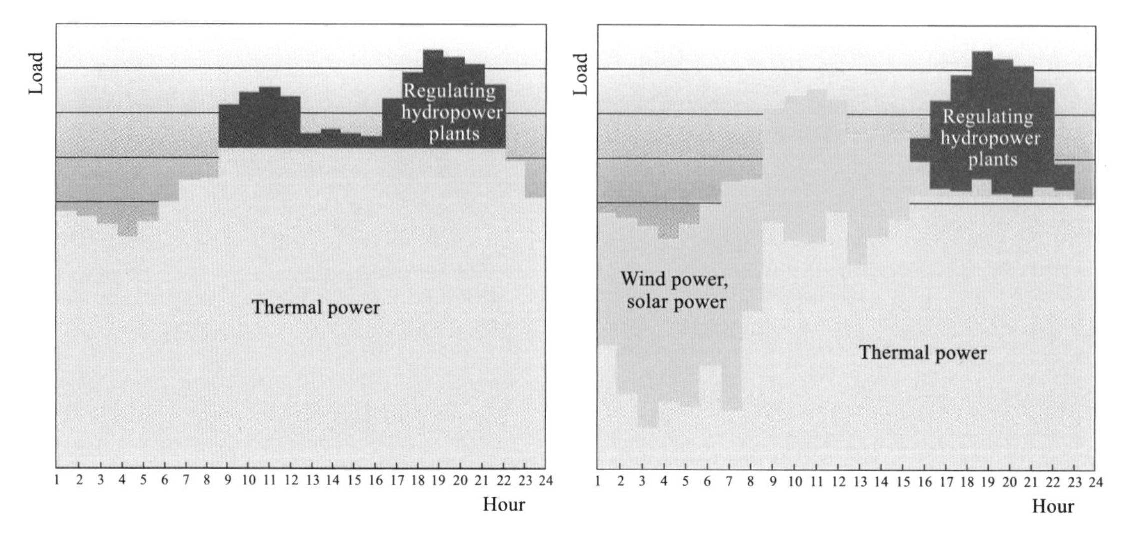

Figure 3. 3. 2 Daily Operation Mode of Hybrid Energy System—Thermal Power and New Energy Power

(3) Construction model.

1) On the basis of the investor's existing thermal power conditions, a solar or wind power plant will be built in the vicinity and connected with the same power transmission system.

2) On the basis of the existing thermal power in grid, a new energy base will be built in the area with abundant resources and favorable development conditions and be connected with the same power grid.

3) The thermal power and comprehensive clean energy base will be planned as a whole, which are bundled and transmitted via the UHV channel to the load center or power consumption areas.

3. 3. 1. 3 Hybrid System for "Pumped Storage Energy-Wind (Solar) Power"

(1) Characteristics.

1) In the local power grid, the pumped storage energy and new energy are mutually complementary. The primary operation mode is power storage in idle time and power generation in peak time that works as the "battery" and "regulating reservoir" of new energy. It helps suppress the impact on power grid caused by unstable output of new energy, enhance the capacity of local power grid accommodating new energy, improve the operational conditions of thermal power units and consolidate the safety, stability and cost-efficiency of the entire electric power system.

2) As the "auxiliary power source at transmission end" of new energy transmission, the pumped storage power plant is built in the vicinity of new energy base to form a hybrid power system platform.

(2) Functions.

1) It reduces the output volatility and protects the safety and stability of power grid.

2) It saves investment in the power transmission system or improves the utilization rate of transmission lines and new energy as well.

3) It promotes the scale-up development of new energy and satisfies the demand of power consumption.

4) It spurs long-distance and interregional consumption of new energy.

Refer to Figure 3. 3. 3 and Figure 3. 3. 4 for the sketch of typical daily operation mode of pumped storage energy and new energy in the power grid and the sketch of operation mode of pumped storage energy in the new energy base respectively.

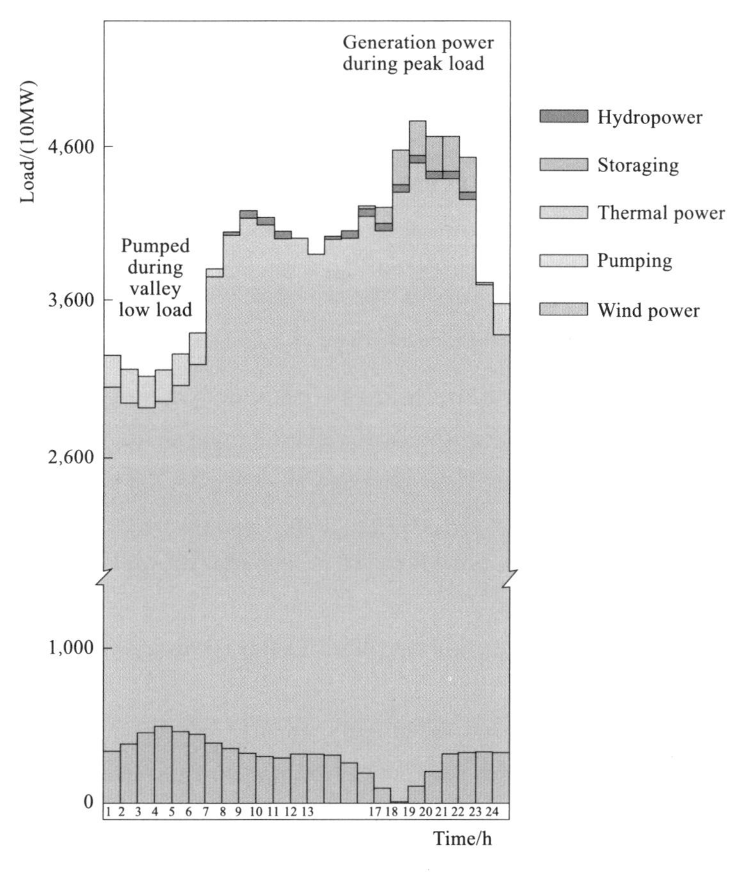

Figure 3. 3. 3 Sketch of Typical Daily Operation Mode of Pumped Storage Energy and New Energy in Power Grid

(3) Construction model.

1) Planning and site selection are exclusively conducted with reasonable layout of pumped storage power plant.

2) As a grid regulating tool, a pumped storage project is generally invested and constructed by a power grid company, within a pumped storage power plant connected to the

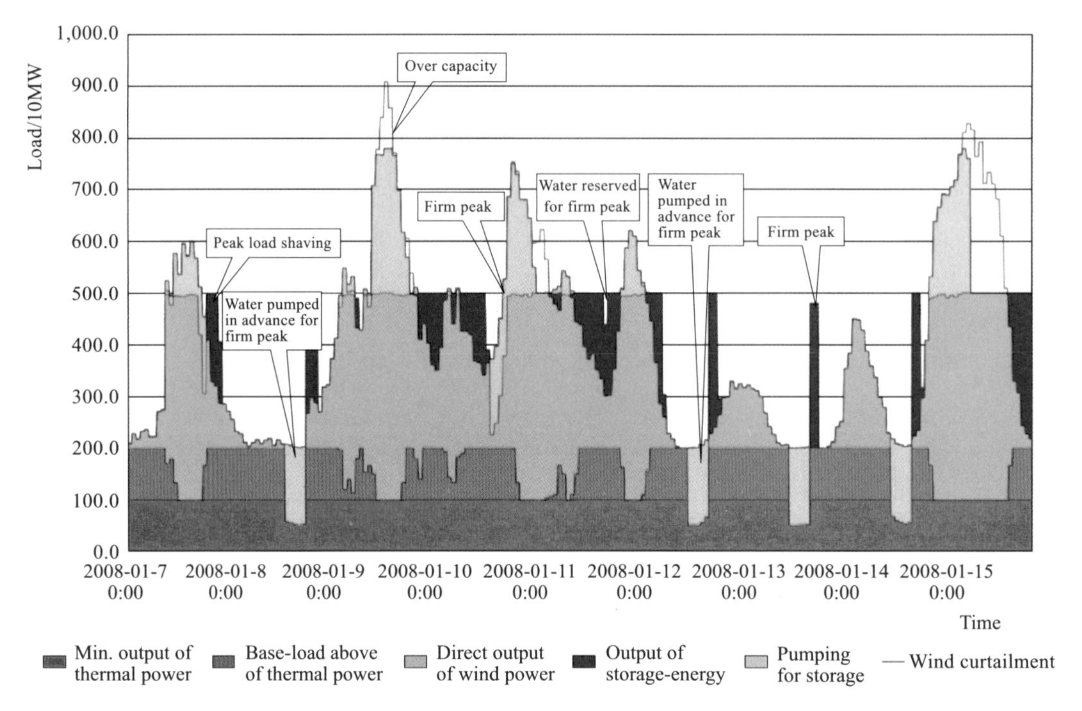

Figure 3.3.4 Sketch of Operation Mode of Pumped Storage Energy in New Energy Base

grid, a new energy base will be built in the area with abundant resources and favorable development conditions and be connected with the same power grid.

3) In the area with abundant new energy, if there is a suitable site for pumped storage energy, the new energy enterprise may consider joint investment in and construction of a pumped storage power plant to expand the development scale of new energy. However, such practice is less cost efficient.

3.3.1.4 Hybrid System of "Energy Storage-Wind (Solar) Power"

(1) Characteristics.

1) The energy storage system can quickly absorb and discharge active power and reactive power that help smooth the new energy output so that the power output per unit time can be controlled within a steady margin, and the new energy output becomes more controllable.

2) The new energy equipped with a large-scale energy storage system is used for peak shaving and valley filling of power grid. In the "valley" period of power consumption, the excessive new energy power is stored, which is then sold to the grid in the "peak" period. The combination of new energy and large-scale energy storage system reduces the burden of peak shaving of the grid, alleviates the supply and demand imbalance of the electric power system and thus improves the economic benefit of new energy.

3) It improves the quality and cost efficiency of distributed power supply.

(2) Functions.

1) It smoothes the new energy output and protects the safety and stability of power grid.

2) It reduces the network loss of system power transmission.

3) It improves the cost efficiency of utilizing new energy.

(3) Construction model.

1) To guarantee the power demand of isolated island or remote area, the off-grid energy storage project is built.

2) To improve the quality and cost efficiency of distributed power supply, the grid-connected energy storage project is built.

3.3.1.5 Hybrid System for Pure New Energy

(1) Hybrid energy system for pure wind power or pure solar power.

Characteristics: Since such power sources have no regulating capability, this kind of hybrid energy system features natural complementation. This complementation does not improve the guaranteed output (guaranteed power generation capacity). This complementation narrows the margins of output; It saves the scale of power transmission.

Functions: It reduces the volatility of new energy output. It saves the investment and construction costs of power transmission system, and facilitates overall management. It improves the utilization rate of new energy.

Construction model: It is planned, constructed and developed as a whole and bundled and connected with the adjacent substation, and then connected into the same power transmission platform.

(2) Hybrid system for wind and solar power.

Characteristics: Since such power sources have no regulating capability, the attribute of complementation is a natural mix of the wind and solar power; This complementation does not improve the guaranteed output. This complementation narrows the margins of output. Under the same power transmission scale, the complementation reduces the curtailment rate of feed-in power where the majority of electric energy can be fed out via the common power transmission lines. In some areas, the mutual complementation occurs more or less throughout the year (the month with more wind power output will have less wind power output). The unused land such as the inter-wind zones in the wind farm can be utilized to install solar power panels, which will increase the land utilization rate.

Functions: It reduces the output volatility. It saves investment in the power transmission system or improves the utilization rate of transmission lines and new energy as well. It cuts down the construction cost and facilitates overall management. It saves land resources.

Construction model: On the basis of existing wind power or solar power project owned by the investor, a solar or wind power plants can be built in the vicinity and connected with the same power transmission system.

3.3.1.6 Integrated Hybrid System

(1) Characteristics. Considering the characteristics of different power sources in a regional grid, it gives full play to the combined advantages of resources like wind energy, solar energy, hydro energy, coal and natural gas to promote effective utilization of diverse energy resources such as wind, solar, hydro, thermal power and energy storage.

(2) Functions.

1) It improves the energy supply and demand coordination ability, and consolidates the comprehensive efficiency of energy system.

2) It spurs the consumption of renewable energy.

(3) Construction model. The integrated hybrid system operation is usually implemented in the large-scale integrated energy bases.

3.3.2 Hybrid Energy Systems at User Side

3.3.2.1 Smart Microgrid

(1) Characteristics. A microgrid refers to a small power plant, distribution and consumption system that consists of distributed power, electrical load, distribution facility, monitoring and protection apparatus, etc. There are two types of microgrids, namely grid-connected type and independent type that can realize self-control and self-contained. The grid-connected microgrid is usually connected with an external grid and capable of on/off-grid shift and independent operation. Such grid-connected microgrid is featured by:

1) Micro. Mainly low voltage level, generally at 35 kV and below, and small system scale where the capacity (maximum power load) is not higher than 20 MW on principle.

2) Clean. The installed capacity of renewable energy accounts for above 50% of total capacity, or the comprehensive energy utilization rate of natural gas multi-generation system is above 70%.

3) Self-contained. It internally has a control system to ensure the power load and independent operation of electrical equipment as well as the abilities of self-balanced operation of power supply and demand and blackout start. When running independently, it can assure the continuous power supply to important loads (not less than 2 h). The annual energy exchange between the microgrid and the external power grid is generally not higher than 50% of the annual power consumption.

4) Friendly. The exchange power and durations between microgrid and external power grid are controllable together with the two-way services for the interconnection grid such as reserve, peak shaving and response at demand side so as to satisfy the quality requirements of consumers and realize friendly interaction with the interconnection grid as well as friendly power consumption by users.

(2) Functions.

1) It is an integral part of a smart grid to achieve the integrated supply of multi energy.

2) It admits the distributed power sources as much as possible.

3) It provides electric energy to the ocean islands, oasis and remote areas not covered by power grid.

(3) Construction model. The smart microgrid of energy is built in line with the development demand of cities, industry and business parks, novel towns and rural areas as well as ocean islands, oasis, etc. In areas like cities, commerce, industry and novel towns, the construction of microgrids based on combined supply system of wind (solar) power and fuel gas is encouraged to improve the comprehensive energy utilization efficiency; in the ocean islands, oasis and remote areas not covered by the power grid, the construction of microgrids by making full use of local natural resources is encouraged in accordance with the local conditions.

3.3.2.2 Combined Cooling, Heating and Power by Natural Gas (CCHP)

(1) Characteristics.

1) Cascade utilization: the natural gas is used as the major fuel to drive the equipment like gas turbines, gas internal combustion engines and micro gas turbines, and the high-temperature steam after gas combustion is firstly used to generate power, the moderate-temperature steam after power generation is used for heating or cooling while the hot water thereafter is used for domestic purposes.

2) Utilization based on hybrid system: CCHP can realize coordinated operation with the power grid and other distributed micro energy (wind power, solar, solar energy collector, etc.) plants in the locality can realize peak shaving, valley filling, and scale-up utilization of new energy.

(2) Functions. It is a kind of distributed power that improves energy utilization efficiency, realizes cascade energy utilization and uplifts the primary energy utilization efficiency to above 80%.

(3) Construction model.

1) The CCHP center is usually built in relatively large areas such as industry, commerce and sci-tech parks where the unit of larger capacity is applied.

2) The CCHP system is also used in the buildings of special purposes, e.g. office building, business building, hospital and some other complex buildings where the units of relatively small capacity are applied.

Chapter 4

Practices of Hybrid Energy System

4.1 Case of Hydro-Solar Hybrid Energy System—Longyangxia Hydro-Solar Hybrid Power Project, the World Largest

4.1.1 Project Overview

4.1.1.1 Overview of Longyangxia Hydropower Plant

Located on the conjunction of Gonghe County and Guinan County of Qinghai Province, Longyangxia Hydropower plant is the leading power plant in the cascade development of Long-Qing section of the Yellow River in China. The power plant is situated at the entrance of Longyangxia Gorge. With a normal pool level of 2,600 m, it has the total capacity of 24.7 billion m^3 and the regulating storage capacity of 19.35 billion m^3 with good over-year regulating capability.

The power plant has 4×320 MW turbine generator units and their total installed capacity is 1,280 MW. The 330 kV HV side has 4 circuits of feed-in lines and 5 circuits of feed-outlines and 1 spare circuit. The main electrical connections are double-bus bar single-section feed-out lines with bypass disconnectors.

It is a large-scale multipurpose project with over-year regulating capability. It is featured with a huge reservoir capacity, strong compensation capability, and the operating and dispatching pattern of "power depending on water and water allocated by power", which has the advantageous conditions of the existing reservoir capacity and dispatching operation for compensation and regulation of solar power plants.

4.1.1.2 Overview of Solar Energy Resource

Qinghai Province lies at the region of middle and high latitudes and enjoys strong solar irradiation and long irradiation time. The average irradiation per annum may range between 5,560 MJ/m^2 and 7,400 MJ/m^2 where the direct irradiation amount accounts for more than 60% of total irradiation, ranking No. 2 in China only after Tibet. For the space distribution, Qinghai has more irradiation in the northwestern part than that in the southeastern part. In Hainan Prefecture where the project is located at, the total solar irradiation per annum ranges between 6,200 MJ/m^2 and 6,800 MJ/m^2.

The project is in Gonghe County of northeastern Qinghai Province on the Qinghai-Tibetan Plateau, which is topographically higher in the northwest and lower in the southeast. This mountainous region includes South Mountain, Sun & Moon Mountain, Ela Mountain, etc. Qinghai Province has a climate of plateau sub-frigid zone. The whole Gonghe County area has sufficient and strong solar irradiation with the average daily solar irradiation duration of 8 hours, average yearly irradiation duration of 2,719 hours and average total irradiation volume of 6381.6 MJ/m^2 over the years. On the contrary to the abundant solar energy resource on vast land, Hainan Prefecture of Qinghai Province is under-populated. The planned solar power parks are mostly built on desertification land, which has natural advantage for the construction of solar power plants.

4.1.1.3 Determination of Scale of Solar Power Plant Connected with Longyangxia Hydropower Plant

Longyangxia Hydropower Plant was put into operation and started to generate power in 1987. The rated current is 1,600 A, the rated current of 330 kV bus is 2,500 A, and 2×400 mm^2 conductors for transmission lines.

The connection of the solar power plant should not only ensure the capacity of the solar power plant, but also protect the safe and stable operation of Longyangxia Hydropower Plant. Based on analysis of the transmission capacity and 330 kV electrical equipment, the scale of the solar power plant was determined to be 850 MWp.

4.1.1.4 Overview of Hydro and Solar Hybrid Power Project

The solar power plant is located in Solar Power Park at Gonghe County, Hainan Prefecture, Qinghai Province, which is about 12 km from the county seat and about 30 km of linear distance from Longyangxia Hydropower Plant. Its installed capacity upon the final phase project is 850 MWp with the annual utilization duration of 1,556 hours. The project occupies a land area around 24 km^2. This solar power plant feeds power into Longyangxia Hydropower Plant by one 330 kV circuit and makes use of the existing 5 circuits of transmission lines of this hydropower plant to realize connection with the power grid system.

The hydro-solar hybrid power project is connected to the spare feed-in bay of Longyangxia Hydropower Station by 54 km-long 330 kV overhead transmission lines and through regulating by water turbine generator units, it delivers the regulated power to the power grid via the delivery channel of Longyangxia Hydropower Station.

Solar power generation is featured with randomness, intermittence and periodicity. The power plant only works in daytime with zero output at night; when it generates power, other power sources in the power grid need to be adjusted to give up the load to solar power supply; when the power plant is blocked out by cloud, its output may plunge while other power sources should immediately react with larger output to fill up the gap caused by reduced solar power supply. Decided by such natural characteristics, in grid-connected operation the active power output of the solar power plant must be compensated and regulated by other regular power sources. The fluctuation of solar power output is absolutely dependent on random change of weather conditions, which is more volatile compared with the normal load fluctuation of the power grid. Therefore, the regulating capacity prepared for solar power generation should be provided by standby rotating generation equipment instead of temporary start/stop of units.

The complementary and coordinative operation of hydro and solar power can maximize the energy utilization efficiency and satisfy the power dispatching requirement of grid. Provided that the downstream water demand is guaranteed, it may to the greatest extent ensure the solar power supply by mitigating the negative effect of bad weather, and is relatively safer. The power generation characteristics of hydro energy and solar energy can be utilized to make up the disadvantages of an independent solar power plant and guarantee the stability and reliability of power sources. Thanks to the fast regulating capability of turbine generator units and the regulating competency of Longyangxia Reservoir, the active power output of the solar power plant can be regulated to produce a smooth solar power output curve and better power quality.

After the project completion, the solar power plant and hydropower plant are coordinated in operation as the same power source point. Based on the water volume balance of the hydropower plant, it makes use of the regulating capacity of the reservoir and the quick regulating performance of hydropower units that to the greatest extent guarantees the solar power output, realizes smooth output of internal power sources and thus provides more friendly power to the grid. In the meantime, the solar power plant serves as the virtual hydropower unit for non-loss energy storage and generation that better utilizes the clean energy such as hydro and solar energy resources.

4.1.1.5 Project Construction

Longyangxia hydro-solar Hybrid Power Project includes the Phase Ⅰ 320 MWp solar PV power project and the final capacity after the project completion reached 850 MWp, which was the world largest centralized grid-connected solar power project at that time and the first grid-connected solar power plant in the world adopting the hydro-solar hybrid mode to coordinate and operate power generation.

In line with the design purpose of Longyangxia Hydropower Plant and its development, the Longyangxia hydro-solar hybrid power project is defined as mainly supplying power to Qinghai Power Grid. In December 2013 the Phase Ⅰ 320 MWp project passed acceptance check and was connected to the power grid. In November 2015 the whole 850 MWp project was put into operation.

4.1.2 Key Technology

4.1.2.1 Research Based on Hydro-solar Hybrid Power Project

When the project was being constructed, the large-scale centralized grid-connected solar power plant was in its infant stage in China. It was for the first time to apply the combined regulating and control of hydropower units and solar power generation with little experience for reference. Plenty of researches have been conducted on this project. Taking into account the characteristics of Longyangxia hydro-solar hybrid power project, the regulating characteristics of the turbine generators, governors and SCADA system, as well as the functionalities of the solar power forecast system, the centralized monitoring system of the solar power plant, the coordinative operation, mutual complementation of the hydropower, solar power plants, and the regulating capability of hydropower units on different scales of solar power generation are studied, so as to realize the AGC and AVC functions of coordinative operation control between hydropower and solar power plants. Main researches focused on the analysis of light resource characteristics and power generation characteristics at Gonghe County of Qinghai Province, analysis of hydro-solar hybrid power generation, analysis of hydro-solar hybrid operation impact on the power grid, study of the grid-connected solar power plant connectivity with Longyangxia Hydropower Plant, study of the grid-connected solar power plant in coordinative operation based on the hydro-solar hybrid power system, the control scheme and test program for coordinative operation of the hydro-solar hybrid power system.

These researches determine the control methods and regulating principles of coordinative operation of the hydro-solar hybrid power system, propose a set of control schemes suitable to the coordinative operation of a large-scale grid-connected solar power plant and a large hydropower plant, improve the energy quality of solar power, and attain the goal of utilizing solar energy and hydro energy as much as possible.

The researches theoretically demonstrate the technical feasibility of hydro-solar hybrid power system and propose the philosophy of coordinative development of hydro-solar hybrid energy system that creates favorable conditions for further promoting the coordination technology of hydro-solar hybrid energy system.

The solar power plant and the hydropower plant serve as the same power source point. In consideration of water volume balance, the control principle is determined to try to assure guarantee the solar power output.

In line with the reservoir regulating capacity, the project makes use of the quick regulating performance of hydropower units to optimize the solar power output curve and provide quality electric energy to the power grid. The overall output curve after hydro-solar complementation can satisfy the grid's requirement on volatility margin and strengthen the grid operation stability.

Compared with an independent solar power plant, the solar power plant regulated by a hydropower plant is more friendly to the electric power system and reduces the spinning reserve capacity required by the power system to accommodate an independent solar power plant.

The hydro-solar hybrid power system turns the grid-connected solar power plant requiring all-grid active and reactive compensation into the local hydropower plant compensation, and thus improves the safety and stability of electric power system.

On the basis of not affecting the power generation benefits of the hydropower plant, it adds extra annual utilization hours to the existing transmission project, and makes the grid construction investment more cost efficient.

It proposes a whole set of research theories on coordinative operation of hydro-solar hybrid energy system and its control strategy and functions, develops the automatic generation control (AGC) and automatic voltage control (AVC) software for the control of hydro-solar hybrid operation that is applied to the monitoring system of its coordinative operation to realize smooth output of intermittent power source and improve power grid safety and stability; it also testifies the hydro-solar hybrid control strategy and software that provides the theory and practice basis for the future coordinative operation of hydro-solar hybrid energy system.

4.1.2.2 Integration and Optimization

In recent years, China strongly pushed forward the development of clean energy. The National Energy Administration (NEA) has issued quite a few policies to speed up the solar PV industry development. Qinghai Province also makes use of its excellent solar resource and the advantage of available land resource to develop solar PV industry. Thanks to the strong support of NEA and Qinghai Province, power source enterprises are also actively engaged in the development and construction of solar PV projects, and build up a complete industry chain of well-established management system that ranges from design, research and development to practice demonstration.

System integration is of utmost importance in the construction of large-sized solar power plants. By improving the technical parameters and efficiency of equipment, the project consolidates the overall technology level of the power plant under the standard engineering management, and by applying the reasonably designed and optimized new technology of solar industry, the system integration and transformation efficiency of the entire solar power plant are improved so as to decrease the power generation cost.

4.1.3 Project Characteristics

4.1.3.1 Characteristics of Power Generation of Large-scale Solar Power Pant

At the region of Gonghe, the average daily power generation within a month is relatively stable. The characteristics of daily change are mainly decided by seasonal factor, daytime and nighttime, weather, temperature, etc. The daily output changes of a solar power plant are more or less volatile, random and intermittent. In sunny days, the output curve is a smooth downward parabolic curve while in other weather conditions, the output curves are zigzagging with different volatility margins.

4.1.3.2 Operation Characteristics of Hydro-Solar Hybrid Power System

From the perspective of the grid connection system, Longyangxia hydropower plant and the solar power plant is deemed as one power source point, i.e. when the daily dispatching authority designates the overall power target of this hydro-solar hybrid power project, the new power generation plan assigned to this project is developed on the previous basis of daily power generation plan of Longyangxia Hydropower Plant with the overall consideration of cascade power generation and downstream water consumption demand. Given the power generation characteristics of a solar power plant, the overall daily power generation plan is made after addition of the solar power output. The AGC and AVC control of Longyangxia hydropower units and the solar power plant are conducted through the hydro-solar hybrid coordinative operation control system. The solar power plant and Longyangxia hydropower plant are combined together to receive the dispatching orders of the electric power system and participate in the automatic generation control of the power grid.

From the perspective of regulating means of coordinative operation of the hydro-solar hybrid system, the solar power plant and Longyangxia Hydropower Plant are deemed as one power source point under the system dispatching and regulating, while from the perspective of output, the solar power plant may be deemed as the expanded #5 unit of Longyangxia Hydropower Plant. However, compared with the other four units, it has the following characteristics:

(1) In principle, it is not capable of regulating, i.e. with no regulating of active power.

(2) $\cos\varphi=1$。

(3) It generates power but concurrently has no discharge flow.

(4) Its power generation is volatile, i.e. with output at daytime but no output at nighttime so that in normal conditions its daytime output shows a downward curve.

(5) After the coordinative operation of Longyangxia Hydro-Solar Hybrid Power Plant and provided that the downstream water consumption demand is satisfied, the assigned power generation plan is to store the water volume corresponding to the required power output in Longyangxia Hydropower Plant during solar power generation, which is a non-loss energy storage process. In other words, by using Longyangxia reservoir, the solar power plant works as a virtual turbine generator unit that to certain extent enlarges the hydropower capacity in the power grid.

4.1.3.3 Coordinative Control of Hydro-Solar Hybrid Power System

Regarded as the #5 power generating unit, the solar power plant together with the rest four turbine generating units of Longyangxia Hydropower Plant are subject to the dispatching and management orders of the power system, i. e. the power generation plans assigned to Longyangxia Hydropower Plant should concurrently cover all these five units, which have the following features:

(1) The discharge flow is estimated based on the capacity of four units.

(2) The increase/decrease of reserve capacity is estimated based on the capacity of four units.

(3) The power output is estimated based on the capacity of five units, but one unit (solar power plant) has unstable output that changes with time.

Therefore, the daily power generation plan assigned to Longyangxia Hydropower Plant generally considers the sum of power output from the discharge water and output of the solar power plant as well as the time features and load distribution of solar power generation, which can guarantee the full-extent functions of Longyangxia Hydropower Plant among the cascade hydropower plants along the Yellow River and meanwhile guarantee the power generation of new energy as much as possible.

Longyangxia Hydro-Solar Hybrid Power Plant operates in the pattern of coordinative operation based on hydro-solar hybrid system, and a coordinative hydro-solar hybrid power control system based on the monitoring system of Longyangxia Hydropower Plant. It has an unattended (less attended) and open hierarchical all-distributed structure, i. e. by using function distribution method and distributed database system, the information of the solar power plant is connected as one generating unit of the hydropower plant. The AGC and AVC control strategy for hydro-solar hybrid system is developed together with the assumptions of regulating algorithm of multi-unit vibration zone and the regulating strategy of adjustable threshold as well as the grouped control of two or above hydropower units, which not only achieve the full capacity regulating of the solar power plant, but also guarantee the safe and stable operation of hydropower units.

After receiving the power generation order assigned by the electric power system, the coordinative hydro-solar hybrid power control system controls Longyangxia hydropower units and solar power units in the AGC and AVC techniques. When the power generated from new energy is inconsistent with the planned output due to the causes like weather, it can be compensated by the hydropower units of Longyangxia Hydropower Plant, so that the hydro-solar hybrid power output can best satisfy the requirements specified in the power generation plan. If the discharged water volume is influenced by such compensation to solar power output by turbine generator units, the hydropower plant may adjust the reservoir outlet flow by adjustment of power generation plans of the next day or next period.

4.1.3.4 Uniqueness of Longyangxia Hydro-Solar Hybrid Power Plant Compared with Other Solar Power Plants

(1) Different grid connection mode. The hydro-solar hybrid power plant is close to

the hydropower plant. The solar power plant connected with the spare bay of Longyangxia Hydropower Plant operates under the coordinative control mode of the hydro-solar hybrid power system, and makes use of the existing transmission lines of the hydropower plant for connection with the electric power system. Other solar power plants are generally connected with the electric power system by multiple step-up of the voltage. Since the solar power plants have fewer annual utilization hours, when connected with the power system based on hydro-solar hybrid system, the transmission line utilization rate will be greatly improved.

(2) Different compensation method. Longyangxia Hydro-solar Hybrid Power Plant is connected with the grid after compensation by the hydropower plant for solar power, which is kind of point-to-point compensation. The hydro-solar hybrid power output compensated by hydropower is mainly to meet the demand of the electric power system and only in the case of inadequate compensation by hydropower, the compensation by other power plants in the system is required. Other solar power plants are directly fed into the electric power system via substations, which are mainly compensated by other power plants in the system, i. e. kind of network-to-point compensation.

(3) Different dispatching complexity. Longyangxia Hydro-solar Hybrid Power Plant is connected with the power grid after compensation by both the hydro and solar power plants so that its grid-connected output curve has less sudden alteration compared with other solar power plants directly connected with the power grid. With less demand of sudden regulating, the regulating pressure of the power grid can be alleviated.

(4) Quality of electric energy delivered onto the grid. Owing to the influencing factors like climate and weather change, solar power plants have larger volatility of output curves. Compared with other solar power plants directly connected with the power grid, Longyangxia Hydro-solar Hybrid Power Plant is connected with the power grid after compensation by hydro and solar power plants. The Solar Power Plant and Hydropower Plant of Longyangxia serve as one integrated power source subject to the dispatching orders of the grid and the power generation plans assigned by the grid. The quick regulating performance of the hydropower plant helps regulate the output of the solar power plant, which can smooth the solar power output curve and improve the electric energy quality of solar power connected with the power grid.

(5) Different influence on hydropower plants. Longyangxia Reservoir is an over-year regulating reservoir. Based on coordinative operation of Longyangxia Hydro-Solar Hybrid Power Plant, provided that the downstream water demand is satisfied, the assigned power generation plan is to store the water volume corresponding to the required power output in Longyangxia Hydropower Plant during solar power generation, which is a non-loss energy storage process. In other words, by using Longyangxia reservoir, the solar power plant works as a virtual turbine generator unit that to some extent enlarges the hy-

dropower capacity in the grid.

4.1.3.5 Operation Results

Since its commissioning in 2013, Longyangxia Hydro-Solar Hybrid Power Plant has had satisfactory operation performance. After many tests of active preset closed loop regulating of hydro-solar hybrid operation, load-shedding, load curve tracking, grid integrated commissioning of solar power plant, etc., Longyangxia hydropower units are proven to respond to the remote regulating orders of the power grid, and meanwhile offer continuous compensative regulating against the volatility of solar power generation. The hydro-solar hybrid operation finally reaches the power generation and regulating targets designated by the power grid.

4.2 Case of Hydro-Solar Hybrid Energy System—Bui Hydro-Solar Hybrid Power Project in Ghana

Bui Power Authority (hereinafter referred to as BPA) is the proprietor of Bui Hydropower Plant in Ghana. Established by Ghanaian government in 2007, BPA is the operator of Bui Hydropower Plant, which is a wholly-state-owned enterprise under the Ministry of Power. The BPA founding act in 2007 stipulated that BPA owns the franchise of developing the Black Volta River. The land surrounding the project of Bui Hydropower Plant belongs to BPA. Given the abundant solar energy resources in the vicinity, BPA planned to introduce more investors to jointly develop a solar power plant for hydro-solar mix operation with Bui Hydropower Plant.

In March 2017, BPA was scheduled to use its land to develop a 200 MW hydro-solar hybrid project and invited POWERCHINA International Group Limited (hereinafter referred to as POWERCHINA) to the project planning and engineering. In April 2017, upon BPA's invitation, POWERCHINA organized the site investigation of Bui 200 MW PV project. After the on-site investigation, the experts concluded that the project has good technological and economic indicators, outstanding grid connection conditions and promising prospect of development.

On May 16, 2017, POWERCHINA and BPA signed an exclusive MOU on Bui 200 MW Hydro-Solar Hybrid Power Project.

4.2.1 Project Overview

Bui Hydropower Project was fully commissioned in 2014 for the main purposes of power generation and irrigation. Bui Reservoir has a regulating capacity of 7.72 billion m^3 with over-year regulating capability. Its installed capacity is 400 MW with annual average output of 969 GWh. Featured with large reservoir capacity and strong compensation competency, Bui Hydropower Plant has the existing reservoir for compensation regulating and excellent conditions for dispatching operation of solar power plants.

The intended 200 MW Bui Hydro-solar Hybrid Power Project is located within the scope of the Bui reservoir region. The initial program is that, the project construction period will last 2 years and the operation period will be 25 years, including two phases i. e. the Phase Ⅰ 50 MW project and Phase Ⅱ 150 MW project. The whole 200 MW solar power plant has four circuits of 33 kV transmission lines where two sets of same-tower dual-circuit double-split overhead lines are connected with the 34. 5 kV bus side of the newly-expanded 161 kV substation of Bui Hydropower Plant, and connected with the local power grid via the feed-out channels of this hydropower plant. It was initially expected that after completion of the project, the solar power plant and the hydropower plant under coordinative operation would serve as one power source point. Provided that water volume balance of the hydropower plant is guaranteed, it makes use of the regulating capacity of the reservoir and the quick regulating performance of turbines that to the greatest extent guarantees the solar power output and provides friendly and clean electricity to power grid. See Figure 4. 2. 1.

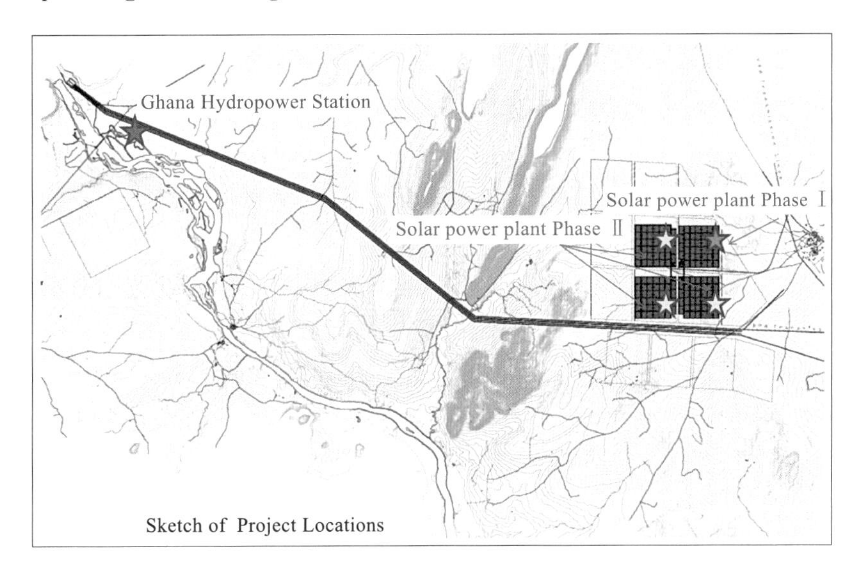

Figure 4. 2. 1 Sketch of Ghana Hydro-solar Hybrid Power Project

As of April 2018, Power China has officially entered the Project Development Agreement (PDA) with BPA, which stipulates the assistance obligation of BPA and the business model. At present, POWERCHINA has launched the negotiation on the PPA of the Project. The Project is scheduled to put into operation at the end of December 2019.

4. 2. 2 Market Circumstances

Ghana mainly depends on thermal power and hydropower. Till the end of 2016, the gross installed capacity throughout the country registered 3,954 MW, where the installed capacity of hydropower was 1,580 MW that accounted for 40% of total capacity; the installed capacity of thermal power (including fuel oil, fuel gas and thermal power) was about 2,351. 5 MW that accounted for 59. 5% of total capacity plus the rest 22. 5 MW of installed capacity of renewable energy; the dependable capacity of power source points reported 3,663 MW.

In 2016, the maximum load of Ghana grid was 2,087 MW and the social power demand reached 13,700 GWh. In accordance with the power generation plan, in 2016 the hydropower plants in Ghana should have produced 4,835 GWh, and the thermal power plants were scheduled to produce 11,566 GWh. However, the thermal power plants failed to achieve the goal due to inadequate fuel supply and the financial problem of Volta River Authority (VRA), the largest power supplier in this country. In 2016, the thermal power output was merely 7,378 GWh, which was 36.2% less than the plan. Confronting the serious shortage of power, Akosombo Hydropower Plant and Bui Hydropower Plant (both are over-year regulating hydropower plants) made use of reservoirs to enlarge output and the actual hydropower output in 2016 amounted to 5,560 GWh, 15% higher than the planned goal. As a result, however, the level of Akosombo Reservoir was lowered to the dead level and the annual minimum level of Bui Reservoir was also 3.78 m lower than the minimum design level. Even so the gross power demand could not be met yet so that Ghana had to import 765 GWh in 2016. At present, Ghana faces severe situation of electric power supply due to lower operation level of hydropower plants, insufficient inflow water and shortage of fuel supply.

Ghana has rich solar resource, but it has not been effectively developed and utilized despite of huge potential. The construction of solar power plant can effectively mitigate the shortage of power supply, and is important to Ghanaian government's vision of 10% proportion of renewable energy in 2020.

4.2.3 Project Advantages

4.2.3.1 Advantages of Construction Conditions

(1) Excellent resource conditions. The distribution of solar energy resources in Ghana is featured with gradual reduction from north to south, i.e. sliding from 2,050 kWh/m^2 to 1,600 kWh/m^2. Pursuant to NASA statistics, the intended site to develop a solar power plant has an average horizontal irradiation of 1,707 kWh/m^2 every year. Pursuant to the standards contained in the *Methods of Evaluating Solar Energy Resource* (QX/T89—2008), the region is classified as a region with "very rich resource" and has considerable value of development.

(2) Developed and reliable power generation and transmission systems. The intended PV project is arranged within the scope of land acquisition of Bui Hydropower Plant that will make use of the existing booster station and transmission lines of Bui Hydropower Plant to deliver electric power. This booster station may well serve as the booster station of the solar power plant, which not only effectively saves investment, but also ensures a developed and reliable power transmission channel.

(3) Advantage in cost. The intended solar PV project occupies the land within the

scope of land acquisition of Bui Hydropower Plant and by in-depth design and integration; it will use part of the common infrastructure of the hydropower plant, which can significantly reduce the civil works cost and further lower the gross cost of the project.

(4) Convenient construction and labor conditions. To satisfy the construction demand of Bui Hydropower Plant, the project owner has built up full and complete temporary facilities including employee camps, canteens, batching plant, carpentry, mechanical workshop, etc. All the temporary facilities may continue functioning during the construction and operation period of the solar PV power plant to spare extra construction cost.

In terms of local employment and labor management, Bui Hydropower Plant during the construction period employed a large number of local workers totaling more than 7,000 people and many of them have worked in Bui project for over three years. The big number of technical workers in principle can guarantee the local labor supply to the solar PV power project construction. In addition, the Chinese management staff had accumulated lots of labor management experience in the construction of Bui Hydropower Plant and established perfect employee performance appraisal and remuneration systems.

4.2.3.2 Advantages of Hydro-solar Hybrid Power System

(1) Advantage of connection mode. The solar power plant for the hydro-solar hybrid system is close to the hydropower plant. The energy output from the solar power plant will be sent via 161 kV substation to the booster station of Bui Hydropower plant, and after mixing hydropower and solar power through the coordinative control system, further sent to the electric power system. Owing to the fewer annual utilization hours of the solar power plant, thanks to the hydro-solar hybrid system for grid connection, the utilization rate of the transmission lines will be improved.

(2) Advantage of compensation mode. The hydro-solar hybrid power plant is connected with the power grid after the compensation between hydropower plant and solar power plant, which is kind of point-to-point compensation. The hydro-solar hybrid power output will be compensated by the hydropower plant to satisfy the power system demand. The system may only need compensation by other power plants in the case of inadequate compensation from the hydropower plant.

(3) Advantage of dispatching complexity. The solar power generation is considerably influenced by weather change. Any sudden change of weather could lead to the sudden alteration of output. Since the hydro-solar hybrid power plant is connected with the power grid after compensation of hydropower and solar power plants, the output curve of the hybrid power plant connected onto the grid will have less sudden volatility compared with that of the solar power plants directly connected to the grid. Less regulating for sudden change helps alleviate the regulating pressure on the power grid.

(4) Advantage of grid-connected electric energy quality. Influenced by factors like

climate and weather change, various solar power plants may experience larger volatility of output curves. Compared with the solar power plants directly connected with the grid, the hydro-solar hybrid power plant is connected with the grid after compensation of hydropower and solar power plants. The solar power plant and hydropower plant are integrated as one power source point that is subject to the grid dispatching and jointly completes the power generation plan assigned by the grid. It makes use of the fast regulating performance of the hydropower plant to regulate the output of the solar power plant and smooth the power generation curve of the solar power plant, which would improve the quality of electric energy fed into the power grid.

(5) Advantage of influence on hydropower plant. Bui Reservoir is an over-year regulating reservoir. After the coordinative operation of hydro-solar hybrid power Plant, provided that the downstream water consumption demand is satisfied, the assigned power generation plan will store the water volume corresponding to the required power output in Bui Hydropower Plant during solar power generation, which is a non-loss energy storage process. In other words, by using Bui Reservoir the solar power plant works as a virtual turbine generator unit that to some certain extent enlarges the hydropower capacity of the power grid.

4. 2. 4 Socio-economic Benefits

4. 2. 4. 1 Mitigate Power Supply Shortage in Ghana

Refrained by energy shortage and frequent closing of import petroleum and natural gas pipelines as well as inadequate funds of Ghanaian government and power authority for new power plants and technical transformation, Ghana is menaced by severe problem of power supply security. The electric power shortage occurs all over the country including its capital Accra where the gaps of government, enterprise and resident power supply are widening. The shortage of electric power turns out to be the biggest administration crisis of the government.

Therefore, Bui Hydro-Solar Hybrid Power Project provides clean energy power. It's predicated that by 2020, the power shortage of the grid in Ghana will reach 1,455 GWh.

4. 2. 4. 2 Optimize the Energy Consumption Structure in Ghana

The present domestic energy development and utilization of Ghana mainly focus on petroleum and natural gas resources. However, facing the energy shortage and increasing market demands, even the huge consumption of petroleum and natural gas can no longer meet the swelling demand. The government plans to improve the utilization rate of renewable energy like solar energy and has worked out many incentive policies to promote new energy market growth, e. g. the *Renewable Energy Act 831* (*REA831*) scheduled to uplift the proportion of renewable energy power generation up to 10% in this nation. Currently, the installed capacity of renewable energy resources (excluding hydropower) is only 22. 5 MW, lower than 1% of the power generation. In addition, the Ghanaian government also proposes a reliable energy market of "diversified and mixed energy" and hopes to form partnership with other countries to improve the operation efficiency of all types of energy and share its energy resources with other countries.

Ghana has rich solar energy resource and promising prospect of solar energy utilization, which complies with the government requirements on developing renewable energy and is beneficial for the diversification of its energy structure.

4.2.4.3 Relieve the Financial Pressure of the Ghanaian Government

According to information, the price is relatively high in the PPA of the oil-fired thermal power recently signed by Ghana. The main reason is that the fuel price is high due to the short supply. According to the PPA on GF thermal power project signed in December 2014, the feed-in tariff including tax is US cent (US₵) 17.2832/kWh when the heavy oil is used as the fuel, and US₵ 12.7516/kWh when the natural gas is used as the fuel. According to the PPA on Astro Power Station signed in December 2015, the feed-in tariff is US₵ 13.1924/kWh when the light oil is used as fuel and US₵ 14.0909/kWh when the natural gas is used as the fuel.

The feed-in tariff of Bui Hydro-solar Hybrid Power Plant will not be higher than US₵ 9/kWh, which has great significance to alleviating the financial pressure of the government.

4.2.4.4 Protect the Ecological Environment of Ghana

Ghana is one of the ten most polluted nations or regions in the world. Every year, environmental destruction, such as water contamination, air pollution, deforestation and desertification, would cause an economic loss of about 10% of its GDP. Its environmental conditions give an alert that its ecosystem carrying capacity is close to the limit. If the proportions of clean energy and renewable energy consumption cannot be increased, its economic and social development might be seriously hampered.

The solar energy is a kind of clean and renewable energy. 200 MW Bui Hydro-solar Hybrid Power Project registered an average power output of 330.7709 GWh per year. Supposing the (standard) coal consumption of thermal power is 315 g/kWh, this project after commissioning can save 104,200 tons of standard coal each year, and accordingly reduce the emission of many air pollutants, e.g. less emission of 324,000 tons of CO_2, 28.36 tons of CO, 1246.53 tons of NO_2, and 1458.96 tons of soot.

4.3 Case of Smart Energy Grid—Regional Renewable Energy Grid Project at Shuanghu County, Tibet

Located in the north Tibet Qiangtang Plateau in the northwestern part of Naqu Prefecture, Tibet Autonomous Region, Shuanghu County is at the buffer area of Qiangtang National Reserve Zone with an average altitude of 5,000 m. It lies in the typical cold semi-arid plateau monsoon climate area with the features of low temperature, large day-night temperature difference, little rainfall, thin air, low air pressure, strong solar irradiation and sufficient sunshine. It has the annual average temperature of −13℃, the average temperature in January of −24℃, the extreme highest temperature of 18℃ and extreme lowest temperature of −33℃. The local annual average sunshine duration is above 3,000 hours and the solar radiation is over 7,000 MJ/m^2. It is the youngest county in China and with the highest elevation. Its land area is about 116,600 km^2. By the end of 2016, the total population of the county is about 15,000.

Before 2015 the county seat of Shuanghu depends on the power supply by 220 kW PV power plant and one 450 kW diesel generator. Owing to the small installed capacity of these power sources, their limited power supply only guaranteed the demand of necessary government agencies and social functions like government offices, health service center, public security bureau and gas station.

4.3.1 Project Overview

The regional renewable energy grid project at Shuanghu County in Tibet is one of the national projects to supply power to those none electrified regions in Tibet. The project targets at Shuanghu County in Naqu Prefecture are to solve the problem of power supply to the government agencies and local residents of surrounding areas. After due consideration of local rich solar energy resources and environmental and meteorological conditions, the projects scheme adopts an independent regional renewable energy grid system based on the energy storage technology, and its output is mainly guaranteed by solar power, energy storage and diesel power generation.

The project construction includes the power source component, including a new solar power plant and an energy storage system, and the transmission and distribution component of reconstructing the power grid system in the county seat. The power source component involves one 13 MW solar power generation system and one 7 MW/23.5 MWh lithium-ion battery storage system which is equipped with two types of lithium-ion battery sets, or namely 3 M/10.08 MWh lithium phosphate battery and 4 M/13.44 MWh ternary lithium battery; the power source system is also equipped with two 1,000 kW/10 kV high-voltage diesel generator sets as the standby power. The urban distribution network involves upgrading of existing 10 kV overhead lines and transformation of 0.4 kV cable lines for residents as well as installation of electric energy meters for every household. The PV power station occupies a land area of 230,000 m^2, of which the management area for plant operation management and power distribution and dispatching management is about 15,000 m^2 including about 9,600 m^2 for the energy storage system.

An energy management system is set to ensure the reliability and stability of off-grid electric power system, so as to realize the coordinated control and operation of the power sources, grid and loads within the system.

The basic design of the project started in March 2016 and its construction started in

May 2016. In mid-October of the same year, the project was commissioned and started to supply power to the county user. After the project completion, it has been managed by the development organization till now.

In accordance with the power supply target set in the project planning stage and the calculation of design scheme, the power supply scope covered about 5,600 people with annual consumption of 1,100 kWh per capita. After completion of the project, on the basis of the energy storage system scale and 90% DOD operation, this energy storage system can provide electric energy of 18 MWh each day. Supposing its sunlight time's power supply accounts for 60% of total energy, the solar-storage-diesel system is capable of providing a maximum electric energy of 30 MWh and its theoretical annual power supply reaches 10.95 GWh. Shown in the output statistics list in 2017 offered by the operation authority, the total power supply in 2017 recorded 7,656,223 kWh and the monthly power supply suddenly grew up since November 2017, which was over 900 MWh per month i.e. over 30 MWh per day, that was mainly caused by the power demand for heating. The power output in December was 941,025 kWh and specifically on December 10 the output recorded 32,994 kWh. The power output did not reach the design target because of different consumption loads of each month but in terms of system's power supply capacity, the target set in the initial stage of project construction has been realized.

4.3.2 Key Technology

The conventional PV power plant and distribution network rely on the support of a grand power grid. An isolated power generation system relies on the coordinative control and operation of the internal power source, loads and grid to guarantee the stability and reliability of the system, and supply electric power in compliance with the requirements on the electric energy quality.

In recent years, China has proactively carried out research on the technology relating to the isolated power generation and supply system, and harvested a lot of achievements in theoretical studies and demonstration projects. China has built up a batch of demonstration projects of distributed power generation and supply systems where the isolated power generation systems are mainly distributed in the remote areas like ocean islands and those with rich renewable energy. East Inner Mongolia Electric Power Co., Ltd. of State Grid and China Electric Power Research Institute of State Grid jointly built the independent wind-solar-diesel-storage power system at Taiping Forestry Station in eastern Inner Mongolia and the grid-connected wind-solar-storage power system at Chen Barag Banner, which successfully solved the electricity shortage in the area. In 2011, China Guodian Corporation built up an island microgrid system involving wind, solar, diesel and energy storage at Dongfushan Island of Zhejiang Province, which was equipped with 100 kWp solar power, 210 kW wind power, 200 kW diesel generator and 960 kWh lead-acid batteries. In 2013, the offshore wind farm company of China Southern Power

Grid deployed a microgrid at Guishan Island of Zhuhai involving 5,000 kW (5×1,000 kW) diesel generators, 2,250 kW (3×750 kW) wind power generators and 4,800 kWh (4×500 kW×2.4 h) energy storage facility. This microgrid on island was completed and commissioned in 2014. In 2013, China Longyuan Power Group Corporation Ltd. established an isolated power generation system of combined operation of hydro-solar-diesel-storage on the Shiquan River in Ali Prefecture of Tibet.

4.3.2.1 Energy Management System of Local area Grid

In order to ensure the safe and stable operation of an independent local area grid system, a set of grid energy management system is required. Based on the analysis of weather forecast, battery energy quantity and real-time load data, this energy management system performs real-time monitoring and dispatching management of power grid, realizes reasonable allocation of grid power and load, and thus guarantees the stability of grid operation.

(1) Independent microgrid energy management. The control of power source and load in the independent microgrid energy management should meet the real time and quick response demand. By taking due consideration of the static operation characteristics and dynamic response characteristics of various power sources in the system, the multi-time-scale optimized dispatching framework is applied to satisfy the information exchange between the power grid and power sources and establish a real-time, controllable and flexible microgrid energy management system framework, which offers practical monitoring system programs and equipment to the microgrid and effectively solves the safety, stability and cost efficiency problems incurred in microgrid application.

(2) Coordinative control technology of microgrid. The typical operation modes of a microgrid are determined in accordance with the grid structure and the power source load distribution and configuration in the system. Through the master-slave control and peer-to-peer control technology, the problem of voltage and frequency stability of the system under different operation modes can be resolved so as to realize the seamless shift between different modes; the system is also to solve the failure recovery problem in the case of serious failure conditions, develop a black start scheme for the microgrid and improve the failure reaction and safe operation competency of the system.

(3) Energy storage multi-unit parallel control technology. The primary inverter control technologies include master-slave control, multi-unit V/F parallel control and droop control technology. By study of overall frame and parallel control policy of these three control technologies, the respective system simulation models have been established for the in-depth study of current sharing and response speed, etc. and for the analysis of application characteristics and problems of these three control technologies.

In engineering application, the multi-unit parallel droop control technology presently has the problem of voltage and frequency recovery. Through success in demonstrative operation of smaller microgrid systems, it has not been successfully applied to a larger microgrid system and thus has higher technological risk. There is certain operation experience of V/F master-slave control technology where one single power source of system takes the V/F function while other power sources are all P/Q sources that guarantee the system voltage and frequency quality. The power source of V/F function must have a higher reserve capacity that can withstand the load volatility regulating in the isolated grid operation. Fortunately, PCS (Power Convert System) equipment technology progress can guarantee the establishment of independent power sources of larger scale.

(4) Structure and functions of energy management system. The off-grid regional grid energy management system is an integrated system of power distribution, microgrid, dispatching, protection and measurement on the basis of exclusive support platform.

The functional structure of system includes:

1) SCADA system of regional grid.

2) Regional grid dispatching and optimizing control (energy management system) (see Figure 4. 3. 1 and Figure 4. 3. 2).

3) Protection information management.

4) Automation of power distribution network.

5) Electric energy measurement and automatic meter reading system.

6) Environment monitoring.

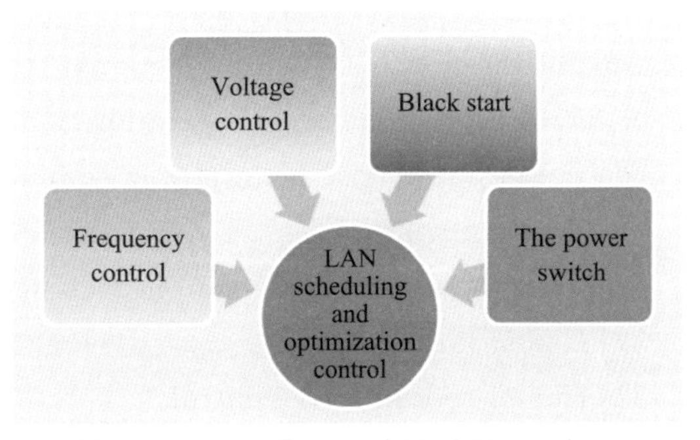

Figure 4. 3. 1 Regional Grid Dispatching and Optimizing Control

The regional grid surveillance and energy management system is the core of coordinated and safe operation of the whole system. The energy management system can realize the advanced application of distribution power sources including black start, off-grid economical operation, off-grid optimized operation, PTUF (Under frequency Load Shedding), over frequency generator tripping, etc.

The energy management system also includes power generation/load forecast, public analysis and optimization of voltage, comprehensive energy efficiency analysis, comprehensive load forecast, smart coordination control, smart data analysis, smart diagnosis and alert, etc.

The core functions of coordinative control of energy management system of the Shuanghu Project are manifested in the following four aspects:

1) System frequency control.

2) System voltage control.

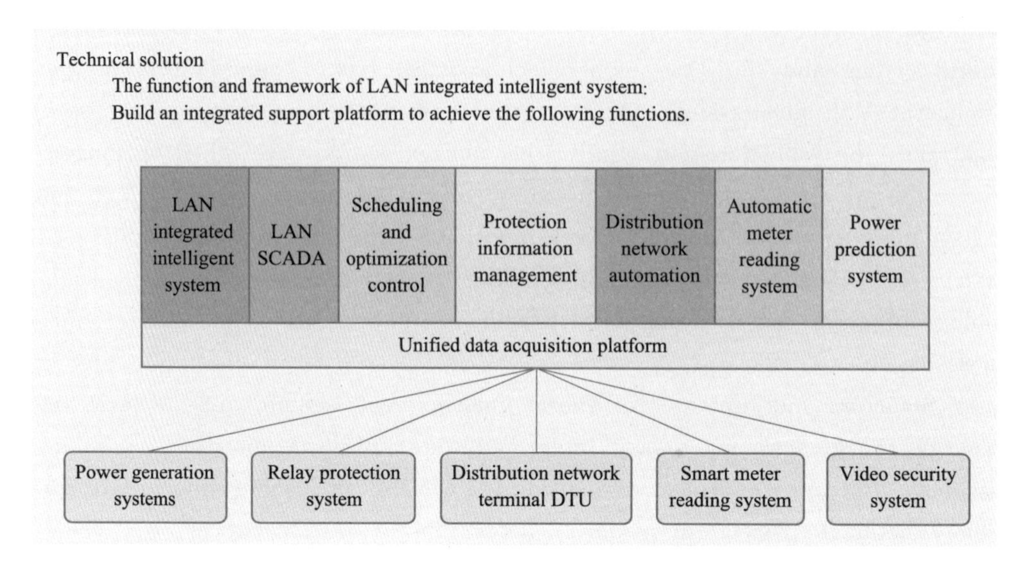

Figure 4.3.2 Energy Management System of Regional Grid

3) Black start.

4) Switch of diesel power generation and main power source of energy storage.

Among these four primary functions, the system frequency and voltage control are the major functions for normal operation of a power plant while the black start and switch between diesel power and main power source of energy storage are auxiliary functions in response to emergency.

Structure of regional grid system includes three tiers of control structure:

1) Tier of local control.

2) Tier of coordinative control.

3) Tier of system control.

Tier of local control: it refers to the control and monitoring devices in system such as switch, circuit, solar power generating unit, energy storage system, distribution transformer, etc. and its main purpose is to coordinate the data collection and control of all units in the system.

On the off-grid run time, through the control model of micro source and setting of control parameters, the coordinative control tier ensures the safe and stable operation of regional grid and maintains the voltage and frequency across the grid within permissible scopes. The coordinative control tier is equipped with two system controllers, one for static control and the other for transient control. The coordinative control tier and on-site control tier are connected by IEC61850 GOOSE to guarantee the real-time system control.

The system control tier performs the comprehensive data collection and processing as well as optimized control and dispatching management of system, guarantees the balance of energy output within the system and also plays other roles like relay protection information management, automation of power distribution, automation of measurement, environment monitoring, etc.

4.3.2.2 Capacity Planning of PV Power Station and Energy Storage System

(1) Load analysis and electricity demand forecast.

The regional grid system load forecast is to analyze the gross demand and the distribution of such energy demand in time so that the appropriate construction scale can be determined in line with the resource characteristics.

The system firstly guarantees the supply and demand balance of gross electricity quantity. Shuanghu County is located at the severe cold area of high elevation and its power consumption is strongly influenced by seasonal factor. Hence the change of power demand in different time periods of a year needs to be considered for output balance.

Therefore, the installed capacity scale of the PV power generation system of the regional grid should satisfy the following two constraint conditions:

1) Day of minimum solar radiant quantity.

2) Day of maximum load.

On the day of minimum solar radiant, the installed capacity of solar power should meet the demand of gross power consumption, or when consumption demand reaches its climax, the solar power output should satisfy the gross demand (see Figure 4.3.3). If the facility is installed with a fixed inclination angle, such angle of modules may influence the output in different seasons. Therefore, the inclination angles of PV modules should be rationally adjusted to uplift the power output in the season of maximum load demand, narrow the gap between power generation and consumption and thus effectively reduce the engineering scale and investment. Through the load analysis and consumption forecast of Shuanghu County, the conclusions are:

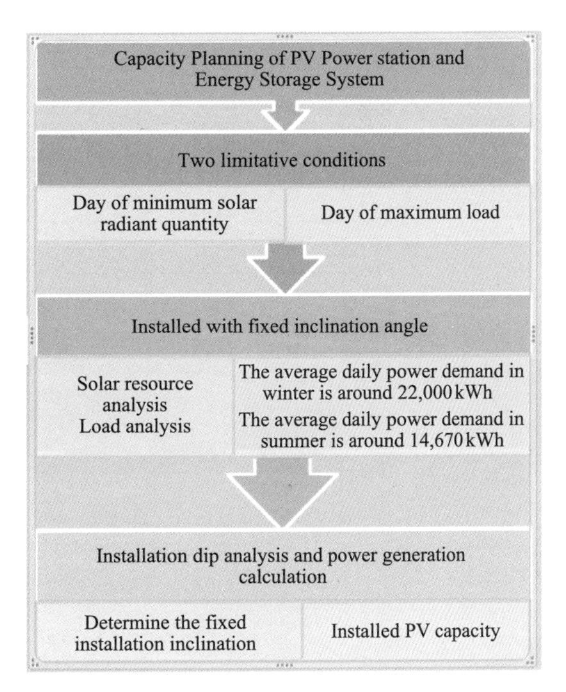

Figure 4.3.3 Determination of Installed Capacity

The average daily power demand in winter is around 22,000 kWh.

The average daily power demand in summer is around 14,670 kWh.

We made use of PV system to analyze the module inclination angle and calculate output. The installed inclination angle of PV modules of the Shuanghu Project should be 54° and the installed capacity should be 11.7 MW. The PV installed capacity should not only ensure the output balance but also satisfy the requirement of energy storage system during the time of less resource. Given the attenuation of PV module capacity year by year, the actual installed capacity of solar power generation system reached 13 MW so that its annual average output may reach 18. 3492 GWh.

Since the installed capacity of solar power system depends on the minimum resource conditions, even when the resource supply is abundant, the output of solar power system cannot be fully utilized.

(2) Determination of energy storage system capacity.

The energy storage system is costly so that its capacity should be determined after careful analysis and study of the system scale. In daytime the power load of regional grid may be directly supplied by solar power, and power supply by the energy storage system is only required in case of nighttime. When the system supplies power to grid load in daytime, the extra output of solar power generation system will be stored in the energy storage system so that in nighttime the power will be entirely supplied by the energy storage system. Therefore, the capacity arrangement of energy storage system should only consider the nighttime power supply. See Figure 4. 3. 4.

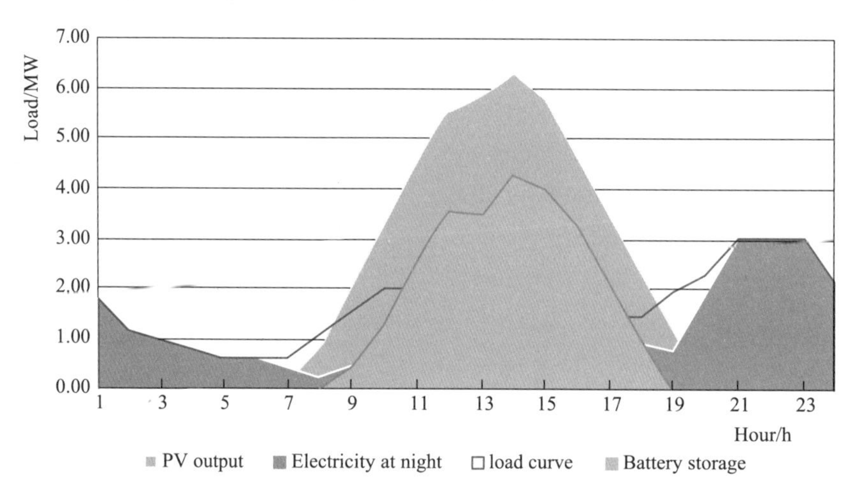

Figure 4. 3. 4 Comparison of PV Power Direct Supply Hours and Energy Storage Supply Hours

The duration of sunshine greatly influences the capacity of energy storage system. In the duration of winter sunshine, the energy storage system has longer time of power supply that requires larger capacity of the energy storage system; in summer with longer sunshine duration, the energy storage system has shorter time of power supply so that smaller capacity of energy storage system can guarantee the power supply (See Figure 4. 3. 5) . The energy storage system needs huge investment. Any irrational expansion of energy storage capacity to guarantee the power demand in winter only results in the incremental investment and such energy storage system cannot be efficiently utilized. Therefore, using diesel generators to fill the gap of power supply in winter can effectively

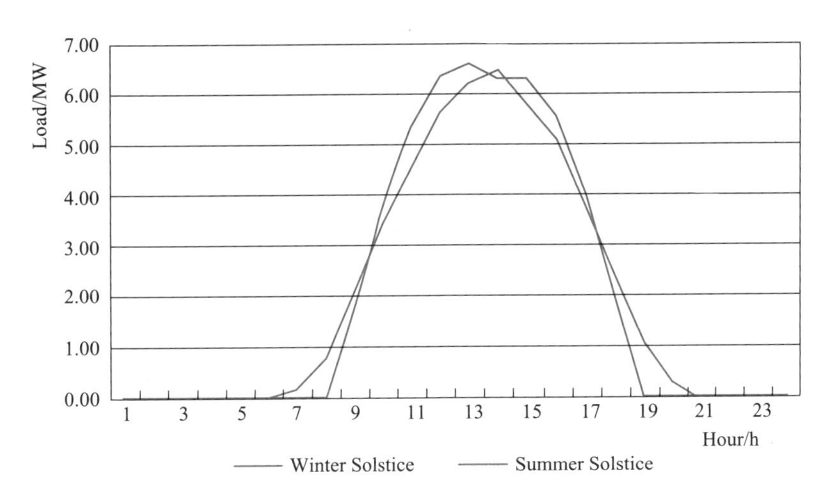

Figure 4. 3. 5 Sunshine Durations on Winter Solstice and Summer Solstice

cut down the investment in energy storage system.

After comparison of different technical schemes, the Shuanghu Project decides the technical scheme for the energy storage system with the total capacity of 7 MW/23. 5 MWh, which is divided into two sets, i. e. one set of 4 MW/13 MWh and the other set of 3 MW/10. 5 MWh.

The two sets of energy storage batteries are boosted by a 3150 kVA 0. 38/10 kV transformer to 10 kV system and then connected with the 10 kV bus bar of the switch station.

The energy storage system is the most critical system of Shuanghu regional grid while the key technology of energy storage system is the parallel operation of maximum eight 500 kW PCS.

The energy storage system of Shuanghu regional power grid can realize the parallel operation of 16 energy storage battery sets, and the key factor to the longevity of the battery system is the consistency of their operation. Moreover, the battery sets are installed in containers. To ensure the compliant operation environment of lithium ion battery sets, the ventilation and heating systems are provided inside containers. By reasonable design of ventilation channels and analysis of interior heat distribution, the temperature is evenly distributed inside the containers so that the battery modules are operating within the allowable temperature range. The longevity of container is another key factor to the service life of the energy storage system.

The multi-unit parallel operation technology imposes higher technical requirements on the design and arrangement of energy storage inverters. There has no precedent of parallel operation of eight energy storage inverters in China so that the regional grid system at Shuanghu becomes an experimental project with certain potential risks. But the present operation demonstrates the equipment meets the anticipated design criteria.

Through the construction of the Shuanghu Project, we have accumulated certain experience in the integrated design and engineering of independent regional power grid system, and have encountered and solved quite a number of unexpected problems during construction.

We have carefully sorted out and summarized the design, installation, commissioning, acceptance inspection, etc. of system and developed valuable technical statistics and standards, which are beneficial to the construction of similar projects.

4.3.3 Project Characteristics

4.3.3.1 Briefing

The regional power grid at Shuanghu is a typical off-grid solar-storage-diesel microgrid system, which is domestically the largest off-grid power system in service. The specially featured environment and climate at the site have rigid technical requirements on all kinds of systems and equipment. By taking reference of the Shuanghu Project, Tibet is building similar projects in other areas.

The off-grid system of regional renewable energy grid at Shuanghu in Tibet may serve as a preferential case for the electric power system of isolated ocean islands. The off-grid system design for isolated islands may be connected with local actual resource endowment and load demand to reasonably plan the power source scheme and the power grid structure e. g. wind-solar-diesel-storage microgrid system so that the types of power source are more diversified and mutually complementary. If needed, certain scale of seawater desalination equipment and cooling equipment may be provided. New energy electric vehicle supply equipment (EVSE) may be installed on the qualified islands so that the electric-driven vehicles on island save the consumption of gasoline and diesel oil. The seawater desalination equipment has adjustable load and can be engaged in the load management in the power system, which is beneficial for the system operation and regulating.

4.3.3.2 Complementary Benefits

In the regional renewable energy grid system at Shuanghu, the solar power generation system is the primary source of output while the energy storage system works as the foundation for the stable operation of the entire power system to set system voltage and frequency. The energy storage system also plays an important role in power output regulating. In daytime with sunshine, in addition to the real-time regulation of power balance between power source and load, the energy storage battery sets with VF function are to uplift SOC to store the electric power as much as possible before sunset; the energy storage battery sets with PQ function work as the load to absorb and store the extra power as quickly as possible under the preset limit. When the external solar power generation system has its output less than the load, the energy management system orders the battery sets with PQ function being converted into the power source, which supplies power as per the preset duty. The V/F source battery sets continue setting the system voltage and frequency and fill the gap of P/Q source. Therefore, the energy storage system is the most critical part of the whole electric power system in terms of time and space balance of energy. The standby diesel oil generators may provide power protection only in consecutive days of cloudy and rainy weather. Such system will make use of the renewable energy as far as possible and minimize the dependency on fossil energy, which is also more cost efficient and environmental friendly to the fragile ecosystem on the plateau.

4.3.3.3 Operation

As mentioned above, the Shuanghu Project is now in smooth operation. The annual power output reached 7,656,223 kWh. The average monthly output was round 500,000 kWh with increase month by month from January to June, which reached 941,025 kWh in December. The maximum daily output of 32,994 kWh occurred on December 10. The power supply capacity of system basically met the design expectation and the power supply capacity also reached its extreme limit. January is the coldest period in a year and the Shuanghu Project stood extreme test. Judging from the power generation result in 2017, we could see that with the commissioning and operation of power plant equipment, the system equipment gradually stepped into stable status with less and less equipment failure, which complied with the trend of system failure probability.

Reflected by the daily load statistics, the load in winter does not vary a lot throughout the 24 hours, basically ranging between 1,200 kW and 1,350 kW with little difference between peak and valley. Because there is almost no industry consumer in Shuanghu and thus the majority of load comes from heating demand. The heating load is relatively stable that helps maintain the stable operation of system.

The result of analysis of PV power system output and charging/discharging of battery system shows that under the irradiation of 10: 00 to 17: 00 each day, besides satisfying the consumption load of the county seat, the system also has excess power to charge the energy storage system. The energy storage system discharges electricity in the rest time that might be as long as 15 hours.

Owing to the high operation cost and low efficiency of diesel generator in high-elevation area, they are rarely triggered as the solar and energy storage system which can basically guarantee the consumption need, which however leads to higher DOD cycling of energy storage system so that the battery SOC is low before the sunrise and may cause breakdown due to the low electric output. It is also negative to the protection of battery longevity.

4.4 The Largest Off-grid Wind-Solar-Diesel Hybrid Power and Seawater Desalination System—Dongfushan Island Microgrid Project in Zhejiang

4.4.1 Project Background

The integrated system of wind-solar-storage-diesel hybrid power and seawater desalination project of Dongfushan Island is located on Dongfushan Island, Putuo District, Zhoushan City, Zhejiang province, the most eastern settled island of Dongji Archipelago in the East China Sea (see Figure 4.4.1 and Figure 4.4.2). Dongfushan Island is the most populated island in the eastern waters of Putuo District with one village (Dongfushan Village) and about 300 permanent inhabitants on it. The local residents live on fishing and working away from home. The island has spiral hilly roads and a ferry terminal. With the Open Sea in the east, it is 45 km away from Shenjiamen Town, Putuo District in the southwest with a land area of 2.95 km^2.

Figure 4.4.1 Rendering Image A of Project

Figure 4.4.2 Rendering Image B of Project

The inhabitants on Dongfushan Island now had a little lighting power supplied by the existing diesel generators on the island. Though the power grid has been laid by the power company, the local dwellers could hardly get power access. The water supply depended on the existing reservoir (capacity of around 10,000 m^3) to collect rainfall for purification and freshwater transport from Zhoushan Islands. The Dongfushan Island integrated system of wind-solar-storage-diesel hybrid power and seawater desalination project mainly tackles with the shortage of water and electricity on the island, which is a public welfare project. It takes advantage of the automated design of coordinated and optimized operation of wind, solar, energy storage and diesel power generation, makes use of the rich renewable clean energy on the island as the primary power source, which is complemented by diesel power generation and equipped with a seawater desalination system to supply freshwater.

With a gross installed capacity of 510 kW, the project covers 7×30 kW wind power units, 100 kWp solar energy unit, 2,000 Ah new energy batteries, 200 kW diesel generator and one set of seawater desalination system with a daily processing capacity of 50 tons, among which the 300 kVA off-grid two-way inverter required by project control is the first one in China. It has a design annual output of 470,000 kWh and annual water supply volume of 18,000 tons. The electric tariff of sustainable project operation is RMB 1.2 Yuan/kWh and the water price is RMB 5 yuan/t.

The feasibility study and investigation design were started in January 2010 and the project scale and design program were determined after several rounds of demonstration; its construction officially started on November 12, 2010. At the end of April 2012, it was completed and commissioned into trial operation.

4.4.2 Design Concept

With complex terrain, Dongfushan Island is dominated by mountain land and has relatively more cliffs and rocks. Several relatively flat valleys on the island are the major residence zones.

Dongofushan Island is named as "the Home of Wind" by local fishermen. According to the estimate, the annual average wind speed is about 7.3 m/s at the height of 50m, annual average wind power density 380 W/m^2 and wind power density grade is Grade 3. Therefore, it is highly worthwhile to develop wind energy resources on the island. The daily changes of local wind speed are relatively small. In general, the wind speed in the morning is slightly low while that from the afternoon to the midnight is relatively high. The changes of wind power density are basically the same as those of the wind speed. The local wind energy resource shows that it has sound development value, the daily changes of wind resources are relatively steady and therefore the wind power output is relatively steady.

Its annual average temperature is 17.5℃. An annual average total radiation (GHI) is 4,860 MJ/m^2 – 4,968 MJ/m^2, Grade C. Therefore, Dongfushan Island is abundant in resources. The changes in the distribution of solar energy resources within a day are relatively great, with a shape of inverted parabola. The time duration with high radiation is from 9 am to 3 pm, and the time duration with the highest radiation is from 11 am to 1 pm. The local solar energy has certain development value with relatively fixed time duration high radiation. Therefore, its power output is relatively steady in the daytime.

Dongfushan Island has relatively abundant wind energy resources and solar energy resources, which can be considered as the power source of the microgrid. The wind power development has a better adaptability to terrain. Wind power can be developed both at plat terrain and mountain land. However, solar energy development has a poor adaptability to terrain. The steep mountain land or northward slopes do not have development conditions. Dongfushan Island has more mountain lands and less flat land and the flat land is mainly used to meet the living requirements of the residents on the island. Therefore, the wind power and solar energy system should be adopted as the main power sources for the new energy microgrid of Dongfushan Island. Considering limited areas for the PV modules on the island, the wind power scale will be greater than the solar PV scale. In addition, diesel generators will be configured as auxiliary and backup power source.

Dongfushan Island off-grid new energy microgrid system is a renewable energy supply system composed of wind power and PV and determines rational storage capacity based on optimal economic efficiency.

The operation scheme for the system is determined according to the wind power, PV output characteristics and local household power consumption. Local residents are mainly fishermen and the degree of electrification is relatively low. The main power consumption period is from 6 pm to 9 pm. The way of power consumption is relatively single and concentrated, mainly for meeting the household demand. The solar PV power supply period is not consistent with the consumption period of residents. However, the daily power supply is steady. Therefore, the solar PV system mainly supplies power to the sea water desalination system. Before dawn and in the morning, the wind power generation is relatively low. Besides the resident electricity consumption, storage battery needs to supply power to the sea water desalination system. Due to the high power generation in the afternoon, the storage battery can be charged besides the power supply to the sea water desalination system. During the electricity consumption peak in the evening, the sea water desalination system maintains the minimum operation conditions, and the wind power and storage battery jointly supply power to residents. When there is insufficient output, the diesel power generation is considered as a complementary measure. At night, wind power as power source is used for the operation of sea water desalination system, and the charging of storage battery. Besides working with the wind-solar hybrid energy system, the diesel generators are also mainly used to supply power independently or together with storage battery to residents under severe weather conditions, such as typhoon.

Local residents use water mainly for three meals a day and clean in the evening. There is a small reservoir on the island, which can be used to adjust the fresh water storage, including purified rain water and desalted sea water. During the dry season, the sea water desalination system can produce 50 tons of fresh water a day, reaching the per capita daily water consumption (0.1515 t) of urban residents in Zhejiang Province.

4.4.3 Engineering Description

As an isolated island microgrid system, the project concentrates on One System & Two Networks where the former refers to one comprehensive energy supply system while the latter refers to one electric network and one water network. They mainly include:

4.4.3.1 Composition of Independent Microgrid

Pursuant to the resource conditions of Dongfushan Island, it studied the proportions of capacity of wind power, PV power and diesel generator (see Figure 4.4.3 and Figure 4.4.4).

Figure 4.4.3 On-site Wind Turbines

Figure 4.4.4 On-site PV Modules

Analyzed the load characteristics of residential power consumption and seawater desalination apparatus.

Studied the battery capacity required for the regulating of microgrid operation;

Researched and developed one 300 kVA two-way inverter (see Figure 4.4.5).

Designed a set of computer monitoring system for energy management and load regulating.

Figure 4.4.5 On-site Energy Storage Batteries

4. 4. 3. 2 Composition of Independent Water Network

A set of seawater desalination system is established to solve the water consumption problem of local inhabitants, and the surplus of wind power output is used by the seawater desalination system to produce freshwater. This system runs 14 hours every day and produces 50 t/d of freshwater. See Figure 4. 4. 6.

Figure 4. 4. 6 Seawater Desalination Facility

4. 4. 3. 3 System Commissioning and Operation

This power generation and supply system is an independent power grid designed under the principle of "unattended or less attended". A SCADA system is provided so that the centralized monitoring system is computer-based. The officers on duty at the control room use the man-machine interface equipment like display, keyboard, mouse, etc. of the operator workstation on the console to complete the monitoring task of all electrical equipment. The computer monitoring system includes the main machine and operator workstation, common programmable logical controller (common PLC cubicle), network equipment, printer, etc.

4. 4. 4 Technological Innovations

The project makes use of the renewable clean energy on the ocean island such as wind energy and solar energy as the primary power sources, and the diesel generator as auxiliary power source to erect a wind-solar hybrid power supply system; it applies the new energy battery and automatic control technology to realize the electric load balance and the voltage and frequency stability of power grid, and makes use of excessive clean energy for seawater desalination.

The project has made breakthrough of the DC grid technology applied in the conventional wind-solar hybrid system and utilized the AC grid technology.

The single unit capacity of most wind turbines both at home and abroad is below 10 kW and few manufacturers are capable of producing and launching the wind turbines with single unit capacity between 30 kW and 100 kW. The project uses seven small wind power generators with single unit capacity of 30 kW that pushes forward the manufacturing level of small wind turbines in China.

The project applies two sets of 1,000 aH new energy batteries of long life and large charging/discharging current; the project developed and innovated the DC/AC dual-way inverter with a large capacity of 300 kVA, which was the first application in the isolated power grid system in China.

It uses automatic control technology and dual-way inverter to achieve the power balance and grid stability of isolated power grid, and via its control system realizes the off-peak operation of seawater desalination system and residential power load.

The project consists of 210 kW wind power generator, 100 kWp solar power generator, 200 kW diesel generator, two sets of 1,000 aH new energy batteries and a seawater desalination apparatus of 50 t/d capacity. With the total installed capacity of 510 kW, it is presently the largest isolated power grid on island with wind-solar-diesel hybrid power and seawater desalination in China.

4.4.5 Project Operation and Application Value

4.4.5.1 Making Full Use of Island Resources to Improve the Living Quality of Residents

With a total investment of RMB 22.16 million yuan, the project has been in service for around six years with satisfactory overall performance. It has brought significant social benefits to Dongfushan Island. The wind-solar-diesel-storage hybrid energy system and sea water desalination system for the project has made full use of the island wind and solar resources, fully considered land constraints and effectively given play to the complementary benefits of renewable energy resources. The project is built mainly to satisfy the demand of power and water consumption by the island inhabitants, instead of a profit-making project. So the electric tariff and water fee are collected only for the sustainable operation of the project, and thus it has significantly improved the living standard of inhabitants on the island.

4.4.5.2 Providing Economic Solutions to Areas without Access to Electricity

According to the preliminary estimate, if 35 kV submarine cables with the minimum section are laid directly from Shenjiamen Town to supply power for residents on Dongfushan Island, the equipment and laying cost will be as high as RMB 54 million yuan, greatly higher than the investment in the solution. In addition, consideration should be given to the two-side step-up substations, island sea water desalination system and submarine operation and maintenance cost and long-distance power transmission and distribution loss. Therefore, the project offers an economic solution to areas without access to electricity.

4.4.5.3 Pushing the Island Infrastructure Upgrading

In addition, the project construction also pushes the upgrading of part infrastructures. For example, the original wharf of the island is a fish pier which only can serve ferry boat and fishing boat. To meet the transport requirements of large mechanical equipment, the wharf was expanded and re-constructed. The new wharf has the ability to serve larger barge, improving the island's external traffic capacity and bringing convenience to the island residents for external communications.

4.4.5.4 Serving as a Model for Enhancing the Electricity Penetration

It is the first off grid water and power supply system in China with the clean energy as main power supply source, which effectively enhance the electricity penetration. The project is of great importance to the comprehensive utilization of clean energy, research of independent microgrid system and development of self-recycling system on islands. It won 2012 Excellent Engineering Design Award of China Electric Power Planning & Engineering Association (CEPPEA).

4.5 The First Wind-Solar-Thermal Hybrid Power System in China—Hami-Zhengzhou UHVDC Wind-Solar-Thermal Power Bundled Transmission Project

4.5.1 Project Overview

Hami, located in the eastern part of Xinjiang Autonomous Region, is at the conjunction between Xinjiang and the Central and East China. Hami is an important energy base in China with abundant coal, wind and solar energy resources. It has excellent development conditions that are favorable to the construction of large-scale energy bases.

Zhengzhou, the capital of Henan Province, is a metropolis in Central China, the comprehensive traffic hub of highway, railway, aviation and communication, and the core municipality of the Central Plain Economic Zone. Henan Province has the largest population among all provinces in China. Though maintaining a fast pace of economic growth, it is in lack of both energy and follow-up development capability. It is menaced by the imbalance of energy supply and demand and thus needs the import of large amount of energy.

Hami region has rich coal resources, but it is not appropriate to feed out thermal power. Because Hami suffers serious shortage of water resources that has been over exploited, building a large thermal power base to feed out electricity becomes inappropriate. In the meantime, it has abundant wind and solar energy resource but the local power grid has limited appetite to consume the wind and PV power with slow progress of development and utilization. Therefore, Hami-Zhengzhou ±800 kV UHVDC Wind-Solar-Thermal Power Bundled Feed-out Project mainly focuses on the transmission of clean energy, and the thermal power is mainly for the purpose of peak regulation.

All designed, manufactured and constructed by China, Hami-Zhengzhou ±800 kV UHVDC Transmission Project is presently the DC project of the largest transmission capacity in the world, and the first large-sized UHV project in China for bundled transmission of thermal, wind and PV power. It starts from Hami Energy Base in Xinjiang and ends at Zhengzhou, Henan Province. Going through six provinces (autonomous regions) i. e. Xinjiang, Gansu, Ningxia, Shaanxi, Shanxi and Henan with a total length of 2,290 km and total investment of around RMB 27.94 billion yuan, the project was completed and commissioned on January 27, 2014.

Hami-Zhengzhou ±800 kV UHVDC wind-solar-thermal bundled power transmission project has a thermal power scale of 5,040 MW to be developed by three entities; it has a wind power scale of 8,000 MW of totally 63 wind power projects to be developed by 39 entities; solar power scale is 1,250 MW of totally 21 solar power projects to be developed by 20 entities.

4.5.2 Technical Route

4.5.2.1 Study on the Supporting Wind Power Projects for UHVDC Wind-Solar-Thermal Bundled Power Transmission Project

According to the statistics of anemometer towers in different regions of Hami, the wind energy resources in all these regions were evaluated. The status quo of wind power projects built and approved in Hami was surveyed. In combination of the wind power development plan, current development of power grid and power transmission line planning for Hami, the "supporting wind power projects layout for UHVDC Wind-Solar-Thermal Bundled Power Transmission Project" was proposed through comparison and analysis.

4.5.2.2 Study on the Development Scale of Wind Power Projects

(1) Analysis on the output characteristics. According to the anemometry data of various regions, multiple development scales are preliminarily proposed. Studies are made on the wind power output characteristics, wind power coincidence factor, economic grid connection rate of Hami under difference development scales, as well as the relations between different wind power development capacities and corresponding grid curtailed electric energy amounts.

(2) Analysis of correlationship and complementationship of wind farms between regions. According to the wind data of different regions, plot the monthly and daily average wind speed graphs and the output process graphs for different wind masts within a year, and analyze the correlationship and mutual complementationship of wind farms between regions.

(3) Propose wind power development scale. Under the condition of meeting the control requirements of UHVDC power transmission channel for wind power output, the development scale with high accumulated electric energy and appropriate curtailment of

wind power is selected, and the "development scale of supporting wind power projects for UHVDC Wind-Solar-Thermal Bundled Power Supply Project" is determined. Related power grid investment should be reduced while rationally curtailing wind power.

4. 5. 2. 3 Study on the Development and Construction Program of Wind Power Projects

Based on the project layout and proposed development scale, multiple combinations development and construction programs are preliminary developed, to respectively analyze the output characteristics, accumulated electric energy, curtailed electric energy and output sustainability of different development combinations and determine the development and construction scheme for the supporting wind power projects for UHVDC Wind-Solar-Thermal Bundled Power Supply Project. See Figure 4. 5. 1, Figure 4. 5. 2 and Figure 4. 5. 3.

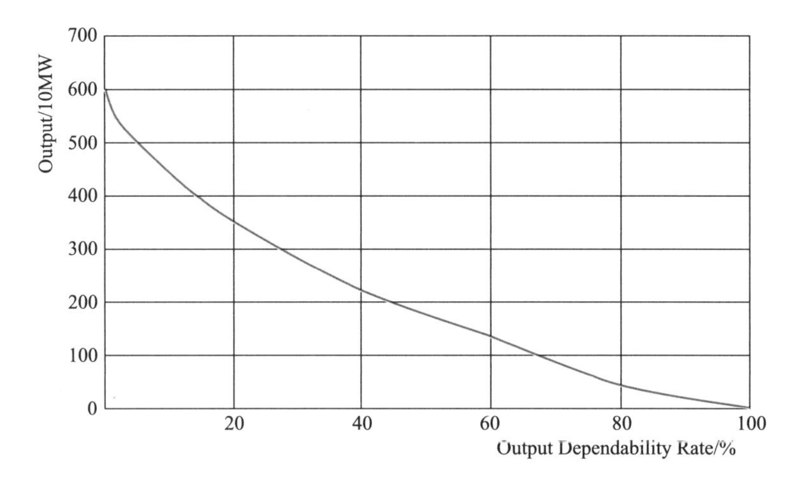

Figure 4. 5. 1 Curve of Output Guarantee Rate

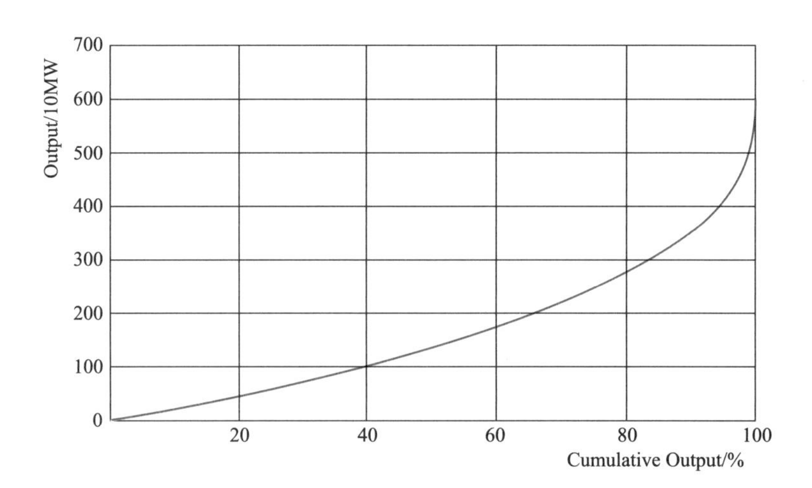

Figure 4. 5. 2 Curve of Accumulated Electric Energy

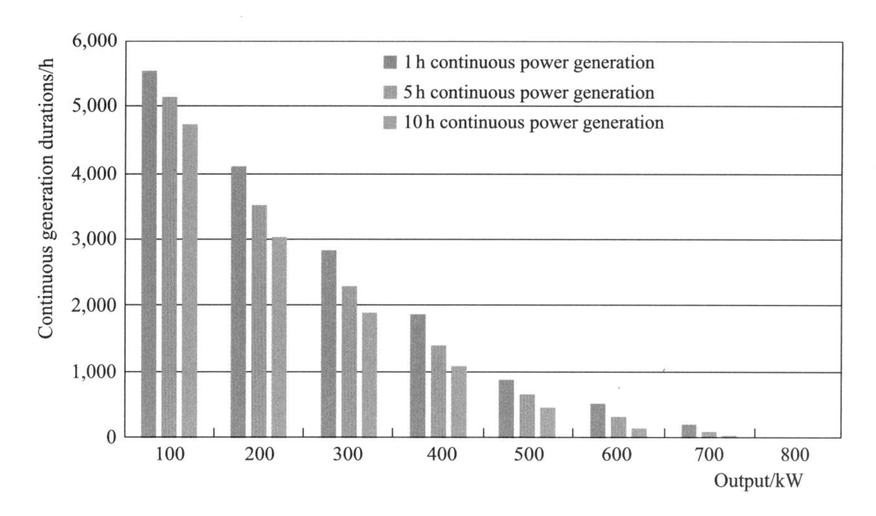

Figure 4. 5. 3　Comparison of Total Power Generation Hours with Different Continuous Generation Durations

4. 5. 2. 4　Feasibility Analysis of Wind-Solar-Thermal Power Bundled Transmission

(1) Analysis on wind-solar power complementation. Mainly analyze the wind power and solar power output complementation on the basis of monthly output and daily output. Preliminarily propose multiple Wind-Solar power installed capacities, make statistics of electric energy transmitted under various power transmission scales and analyze wind-solar hybrid scale structure.

(2) Study on the capacity ratio of wind-solar-thermal power bundled transmission. Make statistics of system frequency out-of-limit, wind power curtailment ratio of system, and thermal power utilization hours based on the project development scale as well as multiple influencing factors, including output change rate of thermal power units, difference coefficient, peak load regulation capacity and channel transmission capacity objectives; analyze the impact on the grid of receiving end and provide the ratio and installed capacity of various energy resources, including wind power, solar power and thermal power, etc.

(3) Economical analysis. Calculate additional generator operation cost, coal consumption cost and operation and maintenance cost of units arising from the frequent peak regulation of thermal power units and convert such cost to the cost increase per kilowatt-hour.

(4) Analysis on technical feasibility. Carry out dynamic simulation of actual thermal power tracking based on the matched wind-solar power output data. Evaluate the feasibility of "Wind-Solar-Thermal bundled transmission" through making statistics of system frequency out-of-limit, system curtailment ratio, and transmission line utilization hours arising from the failure of thermal power units to track due to the rapid decrease of wind power active output under different wind power installation schemes and influencing

factors.

4.5.3 Project Characteristics

Hami-Zhengzhou ±800kV UHVDC Wind-Solar-Thermal Bundled Power Transmission Project is the first key project in China that uses UHVDC lines to supply power. Compared with the existing clean energy projects and those in construction, it utilizes the trans-regional UHVDC power transmission channels for bundled transmission of clean energy and thermal power so as to ensure the complete consumption of clean energy and the stability and reliability of fed-out power.

After completion of the project, it will have a power transmission capacity of 8,000 MW, which can be deemed as the "Silk Road of Power" connecting the western frontier areas with the central plain of China. It assumes the important responsibility of bundled transmission of thermal power, wind power and solar power generated in Xinjiang. Featured with long distance, heavy duty, little loss, environment friendliness and saving land resources, etc., it helps develop and utilize large-scale energy bases, realize the large-scale integrated development of thermal power, improve the development and utilization benefits of energy resources and mitigate the power shortage in Central and East China.

The distribution of energy resource and energy demand is highly uneven, but the important energy bases like thermal power, hydropower, wind power, solar power, etc. are mostly far from the load centers, the programs of hybrid power systems and long-distance interregional power transmission can solve the problem of consuming renewable energy where Hami-Zhengzhou ±800 kV UHVDC Wind-Solar-Thermal Bundled Power Transmission Project may serve as a valuable reference.

For example, in Indonesia, it has the demand for long-distance energy transmission, which offers certain development space for multi-energy bundled transmission so as to improve the energy transmission efficiency and make better use of power transmission channels. In addition, more interconnection of electric power among the ASEAN countries also will create favorable conditions for the bundled transmission of multiple energy resources such as wind-solar-thermal energy and wind-solar-pumped storage, etc., and build a uniformly dispatched and mutually complemented power supply system for the ASEAN countries, which would timely support the power demands of all the member countries, and boost up the overall development of the ASEAN countries.

4.5.4 Complementary Benefits

4.5.4.1 High Utilization of Wind-Solar Power, Rational Curtailment and High Proportion of Clean Energy in the Power Transmission Channel

Hami-Zhengzhou ±800 kV UHVDC Wind-Solar-Thermal Power Bundled Transmission Project has a wind-solar power utilization of 90.2% and curtailed electric energy of 8.35%

for the associated power source. The wind-solar power ratio is 37. 8% of the transmission capacity. At the same time, the utilization hours of power transmission lines are above 5,500 hours. All types of power sources have their economic advantages, and they will not impact the system frequency of Hami power grid (sending end) and Henan power grid (receiving end).

4. 5. 4. 2 Remarkable Energy Saving Effect and Compliance with the Policy of Energy Conservation and Emission Reduction

After put into operation, the project is capable of supplying electric power around 48 TWh/a with a clean energy proportion of 37. 8%, which equals to the saving of about 6. 3million tons of standard coal, and emission reduction of 124,00 tons and 23 million tons of SO_2 and CO_2 respectively.

4. 5. 4. 3 Stimulus to the Local Tourism Development and Economic and Social Progress

The development of Hami-Zhengzhou ±800 kV UHVDC Wind-Solar-Thermal Power Bundled Transmission Project will boost the local relevant sectors such as construction materials, traffic and communication, equipment manufacturing, etc. offer more employment opportunities, develop the tertiary industry and thus trigger and propel the local economic development and social progress.

4. 6 Exploration of Pumped Storage Power Plant, Wind and Solar Hybrid Energy System in Xinjiang—Research on Joint Operation of Pumped Storage Power Plant with New Energy

4. 6. 1 Research Overview

Xinjiang Uygur Autonomous Region, located in the northwestern frontier region of China, boasts wide territory and rich energy resources. In recent years, Xinjiang has witnessed rapid development. The pumped storage power plant site selection planning and the power plant construction are advancing in an orderly way. At the end of 2016, Fukang Pumped Storage Power plant started construction, which is the first in Xinjiang. Xinjiang also actively explores the effective utilization of various energy resources and carried out in-depth researches on the complementary operation of pumped storage power plant and new energy plant.

The operation modes of pumped storage power plant and new energy plants mainly include two types namely the power grid involving pumped storage power plants and the power transmission platform involving pumped storage power plants.

4. 6. 1. 1 Power Grid Involving Pumped Storage Power Plant

By peak shaving and valley filling, the pumped storage power plant serves as the "battery" and "regulating reservoir" of new energy power generation to suppress the impact of unsteady wind and solar power output on the grid, increase the capacity of local grid to accommodate new energy, reduce the start and stop times of thermal power units participating in peak regulating, improve the operation conditions of thermal power units, and consolidate the safety, stability and cost efficiency of the power system. For instance, Fukang Pumped Storage Power plant is located in the city of Changji Prefecture, Xinjiang, about 70km from Fukang City and about 130 km from Urumqi, the load center of Xinjiang. This daily regulating pumped storage power plant has an installed capacity of 1,200 MW, regulating reservoir capacity of around 6.65 million m^3, designed annual energy output of 2.41 TWh, designed annual power consumption for water pumping of 3.213 TWh, and the comprehensive efficiency of 75%. Upon its completion, the project will supply power to Urumqi-Changji Grid. Apart from the functions of system peak shaving, valley filling, frequency regulating, and reserve for emergency case, it may also assist in the consumption of wind and solar power. The aerial view of Fukang Pumped Storage Power Plant construction layout is shown in Figure 4.6.1.

Figure 4.6.1 Aerial View of Fukang Pumped Storage Power Plant Construction Layout

4.6.1.2 Power Transmission Platform Involving Pumped Storage Power Plant

In China, the new energy resources and energy demands are unevenly distributed. The power sources of new energy like wind farms and solar power plants are far from the load centers that require long-distance and heavy-duty transmission. The new energy power generation is featured with centralized development and centralized grid connection. A power transmission platform equipped with a pumped storage power plant of proper scale can utilize the mutual complementation of pumped storage power plant operation and new energy generation, which will make the wind and solar PV power output less random and volatile, improve the cost efficiency of power transmission lines, and reduce its negative influence on the power transmission system, so as to solve the current problems in new energy development and transmission. For instance, Hami Pumped Storage Power plant, located in the city of Hami, Xinjiang, is about 55km from Hami City and is close to the large-sized wind power and thermal power bases of Hami. This daily regulating pumped storage power plant has the installed capacity of 1,200 MW, daily regulating reservoir capacity of about 7.23 million m^3, designed annual energy output of 1.368 TWh, designed annual power consumption for water pumping of 1.823 TWh and the comprehensive efficiency of about 75%. On the Hami feed-out transmission platform, it assists in the transmission of new energy. The aerial view of the Construction Layout of Hami Pumped Storage Power Plant is shown in Figure 4.6.2.

Figure 4.6.2 The Aerial View of the Construction Layout of Hami Pumped Storage Power Plant

4.6.2 Key Technology

4.6.2.1 Key Technology of Pumped Storage Power Plant Involved in Power Grid

(1) Analysis on supply and demand balance of power system. The analysis on the supply and demand balance of the power system is carried out according to the annual load characteristics of the power system design level and various power supplies.

At the feasibility study stage of Fukang Pumped Storage Power plant, based on the load predictions in Xinjiang Grid, the electric power and energy market space is analyzed for Xinjiang Grid and Urumqi-Changji Grid taking into account those power projects built and under construction and those with better preliminary results. The conclusion is drawn as follows: Xinjiang Grid and Urumqi-Changji Grid will have a large electric power and energy market space by 2025.

(2) Analysis on peak load regulating demand of power system. Considering the grid-connection demand of new energy, the calculation for balancing the peak load regulating capacity is conducted to analyze whether the existing power generation equipment in the power system can meet the peak load regulating demand. If not, the possible combination schemes of peaking power sources are studied.

Urumqi and Changji Prefecture are the load centers of Xinjiang. Urumqi-Changji Grid is mainly dominated by thermal power. At the feasibility study stage of Fukang Pumped Storage Power plant, the balance analysis of peak load regulating capacity mainly includes:

Selecting the typical output process of wind power: In order to reflect the impact of wind power grid connection on the peak load regulating of the power system, the typical output process of wind power shall be selected. When the output process of wind power is simulated 10 minutes by 10 minutes in a year, the principles for selecting the output processes of representative design periods are: ①The average output of wind power should be close to that of the series required by the design. ②The maximum, minimum and firm outputs of the representative design periods should respectively equal or close to the maximum, minimum and firm outputs of the series required by design. ③ The representative design periods shall include continuous high wind periods, continuous small wind periods as well as periods equal or close to the period of maximum output variability. ④It should also include those days inconsistent with the load characteristics and those days basically consistent with the load characteristics, and the load characteristics mainly consider the peak periods and valley periods.

Analyzing the output characteristics of wind power to modify the load curve of power system: The statistical analysis is carried out on the output process of Dabancheng Wind Farm per ten minutes in a year, to draw a curve of wind power output-guarantee rate-energy accumulation. Considering the installed capacity of wind power connected to Urumqi-Changji Grid by 2020 is 4,000 MW and in combination of the output process on the typical day, the load characteristic curve of Urumqi-Changji Grid in the typical day is modified.

Calculation of peak load regulating capacity balance and preliminary analysis on peaking power source: The analysis shows that Urumqi-Changji Grid encounters a prominent peaking problem. From the aspect of guaranteeing the safe and economic operation of the power system, the supporting construction of peaking power sources needs to be strengthened. Based on the energy resources of the region, various power sources possibly participating in the peak load regulating are analyzed as follows: ①Since Urumqi-Changji Grid is short of hydropower resources and the surrounding grids do not have abundant peak load regulating ability for hydropower, it is inappropriate to develop conventional hydropower as peaking power sources. ②Based on comprehensive analysis on gas sources, technical economy and industrial policies, it is not suitable to construct gas-fired power plant as peaking power source. ③The peaking load regulating potential of coal-fired thermal power needs to be further exploited. ④To promote the quality of power supply, improve the operating conditions of thermal power units, promote the consumption of new energy, there is absolutely a need for building a batch of pumped storage power plants with excellent peak load regulating performance.

Preliminary analysis on capacity of pumped storage power plants: Based on the power source construction planning and considering such factors as the requirements of peak load regulating of the grid, the installed capacity of pumped storage power plants are preliminarily proposed based on different options, and the calculation is made for the balance of electric power and energy as well as the present value of cost. The following conclusion is drawn: the rational scale of pumped storage power plants in Urumqi-Changji Grid by 2025 should be 1,200 MW to 1,600 MW.

(3) Economic and technical comparison of pumped storage power plant scale. Through the economic and technical comparison, the scale of pumped storage power plant is recommended.

At the feasibility study stage of Fukang Pumped Storage Power Plant, the recommended installed capacity is 1,200 MW considering such factors as geology, project layout pattern, construction condition, electromechanical equipment manufacturing, land acquisition, environmental protection and economic benefits.

4.6.2.2 Key Technologies for the Power Plant Transmission Platform Involving Pumped Storage Power Plant

(1) Analyzing the Output Characteristics of New Energy. According to the design duration output process of wind (solar) power plants, statistics should be made and the curve of output-guarantee rate-energy accumulation should be drawn through analyzing the daily and monthly output characteristics of wind (solar) power, so as to provide a basis for rationally determining the economic grid connection capacity of wind (solar) power. To reflect the randomness and fluctuation, the statistics may also be made for the variability of wind (solar) power output and other indicators according to the project requirements.

At the feasibility study stage of Hami Pumped Storage Power Plant, the output characteristics of new energy were analyzed and the curve of new energy output-guarantee rate-energy accumulation was drawn.

(2) Preliminarily Proposing Supporting Power Source Scheme and Making Calculation for the Joint Operation of Multiple Energy. Based on local resource conditions, a supporting power source scheme is preliminarily proposed; a calculation is made for the joint operation of multiple energy; the delivery power source combination schemes are recommended; the method of present value analysis of cost is adopted to preliminarily

analyze the rational construction scale of pumped water power plants.

At the feasibility study stage of Hami Pumped Storage Power Plant, three combination schemes for delivering power sources of Hami was preliminarily proposed based on local energy resource conditions, development planning and power delivery plan. These schemes include: individual wind power + solar power delivery scheme, wind (solar) power + thermal power delivery scheme, wind (solar) power+thermal power +pumped storage power plant delivery scheme. Through preliminarily analysis, Hami has abundant coal resources and it is planned to build large coal-fired thermal power plants. The individual wind power +solar power delivery scheme cannot reflect the complementary advantages of multiple energy. Therefore, this scheme will not be considered any more. For other two schemes, a calculation was made for the joint operation of multiple energy. The major boundary conditions for the joint operation of multiple energy are as follows:

1) The sum of effective installed capacities of various power sources in each scheme should be less than or equal to the power transmission capacity.

2) Comprehensive peak load regulating rate: according to the technical parameters of coal-fired thermal power units to be adopted, the comprehensive peak load regulating rate of coal-fired thermal power is proposed. In this project calculation, the comprehensive peak load regulating rate of coal-fired thermal power is 50%.

3) Load increasing/decreasing speed of coal-fired thermal power: the load increasing/decreasing speed of coal-fired thermal power per 10 minutes usually should not exceed 10%. To avoid the zigzag change of thermal power unit output in a short time, the operation of thermal power units should be maintained for a while before further increasing or decreasing the load of thermal power units.

4) Coal consumption characteristics: investigate and make statistics of the coal consumption curve of different coal-fired thermal power units, and consider the coal consumption rate of coal-fired thermal power at different load rates.

5) The load rate of power transmission lines: since the power transmission platform has certain adaptability to the receiving-end grid, the capacity of power transmission line is preliminarily restricted at two periods in the calculation for the project. During the valley period, the transmission line is operating at 70% of the maximum transmission capacity; for other periods, the transmission is operating at full load, i. e. 100% of the maximum transmission capacity.

Comprehensive efficiency of pumped storage power plant: the efficiency is proposed according to the layout of power plant, the curve of units operating characteristics and other factors. The efficiency is usually 75% to 80% and the project adopts 75%.

6) Maintenance scheduling: a consideration should be given to the maintenance schedules of different units, such as coal-fired thermal power and pumped storage power

plant.

7) According to the calculation results of multi-energy joint operation, "Wind (solar) power+thermal power+pumped storage power plant" is recommended. The preliminary analysis shows that the rational scale of the supporting pumped storage power plant of Hami Transmission Platform is from 1,200 MW to 1,600 MW based on the calculation of the present value of cost.

(3) Technical and Economic Comparison and Selection of Pumped Storage Power Plant Scale. Through the technical and economic comparison, the pumped storage power plant scale supporting the power delivery platform is recommended.

At the feasibility study of Hami Pumped Storage Power plant, a comprehensive analysis and comparison was made in terms of meeting the requirements of delivery platform, project construction conditions, and economic benefits. It's recommended that the installed capacity of Hami Pumped Storage Power plant is 1,200 MW.

4.6.3 Project Characteristics

4.6.3.1 Project Briefing

Fukang Pumped Storage Power Plant is the first in Xinjiang and also China's first 1,000 MW-scale pumped storage power plant executed in the mode of EPC. The EPC construction was carried out by the lead firm—Power China Northwest Engineering Corporation Limited, associated with Sinohydro Bureau 3 Co., Ltd. and Sinohydro Bureau 15 Co., Ltd.

After the project completion, it will supply power to Urumqi-Changji Grid. The power plant will not only have outstanding performance as a peak regulating power source, i.e. optimizing the grid power source structure, improving the grid voltage level and power supply quality, and guaranteeing the safe and stable operation of grid, but also assist in consumption of new energy that improves the new energy accommodation capacity and cost efficiency of Urumqi-Changji Grid.

Situated adjacent to the Hami 10 GW wind power base, Hami Pumped Storage Power Plant offers service to the power transmission platform of Hami energy base to suppress the wind power output volatility, uplift the new energy consumption capability, guarantee the safe operation of feed-out system and improve the cost efficiency of long-distance power transmission system.

4.6.3.2 Project Progress

The construction of Fukang Pumped Storage Power plant was commenced in the end of 2016 and will be commissioned by January 2024. Hami Pumped Storage Power Plant is presently in the stage of feasibility study.

4.6.4 Project Benefits

While supplying clean power onto the grid, the wind power and solar power are featured with intermittence, randomness and volatility. As the grid connection scale expands, their impact on the electric power system is emerging: to make up for the uncontrollable change of wind (solar) power and maintain the active power balance and frequency stability, other power sources involved in the electric power system should track the wind (solar) power output in a timely way that when the wind (solar) output increases, the regulating power source should lower its output to give up the consumption space to the new energy; when there is less or no wind (solar) power output, the regulating power source should immediately make up the deficit. When the wind (solar) power output is very large while the output of the regulating source in the system cannot be reduced any more, the wind (solar) power will have to be abandoned in order to ensure the safety of the power system. Therefore, to accommodate the consumption of wind (solar) power, other power sources involved in the power grid need to have adequate peak regulating capacity.

The pumped storage power plant is of multi-purpose such as peak shaving, valley filling, frequency and phase regulating and emergency reserve and can apparently promote consumption of new energy, improve operation conditions of thermal power, and ensure the stable and economical operation of power grid, which is a necessary measure to optimize the power source structure and build a smart power grid. The benefits of new energy operation involving the pumped storage power plant mainly include:

4.6.4.1 Operation Benefit of Pumped Storage Power Plant for the New Energy on Grid

By using the energy storage function for energy storage in valley time and power generation in peak time, the pumped storage power plant serves as the "battery" and "regulating reservoir" of new energy to store and convert the wind or solar energy that will stabilize the output of new energy, enlarge the capacity of local grid to accommodate new energy, improve the operation conditions of thermal power units, and consolidate the safety, stability and cost efficiency of power system. Building a pumped storage power plant is of great importance to the scale-up development of wind and solar power within a power grid.

Urumqi and Changji Prefecture are the most developed and active areas in Xinjiang in terms of politics, economy, culture, finance and transportation, which are also the load center of Xinjiang Power Grid. After gird connection of new energy, Urumqi-Changji Grid faces the inadequate capacity of peak regulating. Fukang Pumped Storage Power Plant, once commissioned, will be one of the primary reserve power sources for peak regulating, frequency and phase regulating and emergency cases that reduces the comprehensive thermal power peak regulating rate from 46.5% to 39.0%, and at the same time assists in the consumption of 4,000 MW wind power in Dabancheng and Xiaocaohu within Urumqi-Changji Grid.

4.6.4.2 Operation Benefit of Pumped Storage Power Plant for the Power Transmission Platform

(1) Make the power transmission lines more cost effective. On the UHVDC feed-out

power transmission platform, in order to guarantee the transmission and consumption of wind and solar energy as much as possible, when the wind (solar) power output is larger, the minimum output of thermal power is limited to 50% of installed capacity, which becomes less cost efficient. In particular, when the wind (solar) output is larger in a short time, the thermal power out may frequently change to consistently follow the new energy output, which has negative impact on the operation longevity and cost efficiency of thermal power units.

By serving the power transmission platform, the pumped storage power plant can enlarge the effective capacity of power transmission lines to better accommodate the load change at the receiving end. After serving the Hami Power Transmission Platform, Hami Pumped Storage Power Plant can enlarge the effective capacity of power transmission lines by 1,200 MW, supply corresponding effective capacity to the receiving grid in the peak durations, lower the output capacity in the valley durations, reduce the peak regulating pressure on the grid at receiving end, save the scale of peak regulating power source of receiving grid, and make the power transmission lines more cost efficient.

(2) Impel consumption of new energy. The power transmission platform equipped with certain scale of pumped storage power plant may regulate the wind and solar power operation by "storage of extra output and compensation for shortage" to uplift the wind (solar) power development and utilization rate, enlarge the proportion of renewable energy power in the transmitted electric power, and promote the scale-up development of new energy. The research result shows that the Hami-Zhengzhou ±800 kV UHV DC power transmission project equipped with Hami Pumped Storage Power Plant will raise the wind (solar) power utilization rate close to 93% (up by 3%), and the new energy will account for 38.8% (up by 1%) of the transmitted power that would further promote the new energy consumption.

(3) Protect the safety and stability of power system. The power transmission platform equipped with pumped storage power plant is capable of reducing the wind and solar power output volatility and instant change, mitigating the negative impact of new energy output fluctuation upon the grid safety, and simplifying the complicated operation dispatching the DC power transmission system for higher safety and stability.

Hami of Xinjiang grasps the opportunity of national policy on "Power Supply by Xinjiang" to speed up the pace of grid construction at all levels, cultivate the rigid smart sending end grids, and gradually turn the Hami Power Grid into an energy transmission hub base with cluster generation of thermal, wind and solar power, AC/DC mixed operation and coordinative development of grids at all levels.

With the large scale commissioning of new energy, the proportion of renewable energy power in the gross power supplied by Hami-Zhengzhou ±800 kV UHVDC Power Transmission Project will gradually increase. The volatility of wind (solar) power may impose more or less impact upon the power grid and power transmission platform. Once completed, Hami Pumped Storage Power Plant will improve the safety and stability of the power transmission system, which is of great important to the construction of rigid power grids at the sending end.

4.7 The Largest Whole Underground Natural Gas CCHP Project—CCHP Project of Center Enterprises Headquarters in Shanghai Expo Zone B

4.7.1 Project Overview

For the purposes of creating quality urban space, continuing the utilization of the Expo site in an exemplary way, maintaining the "internationalized, low-carbon and environmental friendly" image of the Expo area, and implementing the philosophy of sustainable development, a Center enterprise headquarters base is built at Shanghai Expo Zone B. The base concentrates on cleanness and efficiency, scale-up development and better environmental quality, positively promotes the utilization of clean energy and follows the principle of low carbon and energy conservation. A natural gas CCHP center is built to meet the energy supply requirements of the headquarters.

The natural gas CCHP system is built on the concept of energy cascade utilization and takes natural gas as the primary energy to produce cooling, heating and power. The system uses internal combustion engine, micro gas turbine and other equipment to produce electric energy, and absorb heat by using lithium bromide units, heat recovery boiler and heat exchangers, etc. so as to realize the supply of multiple energy such as cooling and heating at the user side. The basic process of the system is shown in Figure 4.7.1.

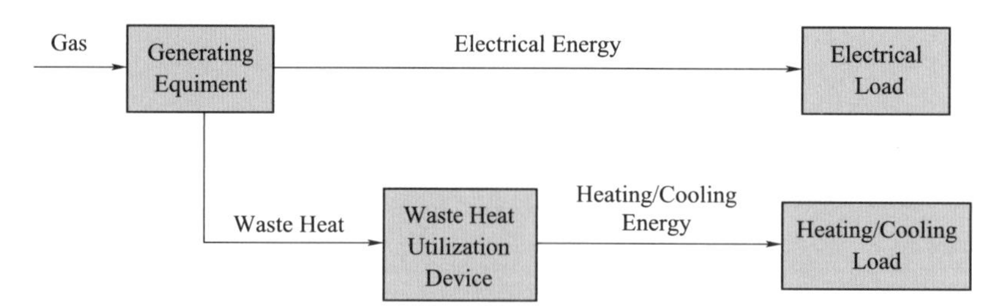

Figure 4.7.1 Basic Flow of Natural Gas CCHP System

Compared with traditional concentrated energy supply, the natural gas CCHP system has the following advantages:

4.7.1.1 High Energy Utilization to Save Energy

The system gives full consideration to the waste-heat utilization. Compared with traditional scattered power supply and heating supply systems, its comprehensive energy utilization can be increased from 40%-50% to 80%-90%, and there is no long-distance transmission loss. The energy conservation rate may reach above 20%.

4.7.1.2 Clean and Environmental Protection

The system uses natural gas as fuel, reducing the emission of SO_2, CO_2 and dust. The reduction rate of CO_2 emission is up to 40%.

4.7.1.3 Fully Meeting the Energy Consumption Demand

The system comprehensively provides three energy, heating, power and cooling, to meet the energy consumption demands of users under various conditions. Upon the system design, different grades of thermal energy including steam and domestic hot water can be provided according to the requirements of users.

4.7.1.4 Improving the Energy Supply Reliability

The system will lower the reliance of the energy supply areas on the grid, ensure the reliable energy supply for important areas and facilities and reduce the load fluctuation of grid.

The CCHP Project of central enterprise headquarters in Zone B, Shanghai Expo ("the energy center") mainly includes: gas engine+flue gas hot water lithium bromide unit, gas-fired hot water boiler, centrifugal cold water unit, thermal and heat storage plant and auxiliary facilities that are arranged under the Power Building of B03 block and Guihua No. 2 Road with a total land area of about 4,200 m^2. The deepest place underground reaches-18 m. The main machine room is of two floors of underground structure and four floors of an underground structure partially. During the design of underground distributed energy of the energy center, full consideration has given to various complex factors of the project. Based on computer-aided analysis, the load simulation and optimization of a distributed energy supply system have been carried out, to build the energy center centered on internal combustion engine-flue gas hot water lithium bromide units. This ensures the reliability of the whole energy supply system, realizes economic benefits to the maximum extent, and enables the well integration of clean energy and green buildings. The commissioning of the project marks the completion of the first whole underground large capacity energy center in Shanghai and has become an example of leading the regional clean energy construction in Shanghai and even in China.

4.7.2 Key Technology

The Central Enterprise Headquarters energy center in World Expo Zone B in Shanghai mainly provides service to the office buildings of the headquarters of enterprises directly affiliated to the Central Government and the subsidiary business services. In accordance with the thermal load capacity and characteristics of Central Enterprise Headquarters in World Expo Zone B, a natural gas distributed energy center is built under Guihua No. 2 Road to supply both cooling and heating service to this area. The project applies novel, energy saving and environmental friendly technology and materials, fully considers the exterior conditions, engineering difficulty and system feasibility, compares and optimizes the engineering options that makes the system more efficient, safer and cost-saving. In particular, the project makes many optimization and innovation, and breaks bottlenecks such as the optimization of cooling storage mode, selection of power supply network, optimization of gas engine cooling plan, type selection and optimization of primary equipment, utilization of pump energy conservation technology, optimization of lithium bromide CCHP technology, optimization of safety and reliability of cooling tower system, which are important to the final project implementation. In addition, BIM three-dimensional design software is applied for collision simulation of equipment and pipelines to consolidate the engineering safety. The primary critical technologies are as follows:

4.7.2.1 Application of CCHP Technology with High Comprehensive Energy Efficiency

The system improves the comprehensive utilization rate of fuel gas. The total generation CCHP of efficiency (35%-45%) and residual heat utilization (35%-45%) is more than 70%. The CCHP satisfies the basic load, and the gas boiler and centrifugal refrigerator regulate the load so that the whole system meets the maximum cooling and heating demand of users, guarantees the stable operation of distributed energy and makes it more cost efficiency than the municipal fuel gas supply. The internal combustion engines applied in the project have higher power generation efficiency and less emission with lower gas inlet pressure. It is featured with less volatility of fuel price, higher equipment integration and fast and convenient installation. The system is equipped with the most suitable type of lithium bromide unit as per the characteristics of selected prime move where the exhaust heat source drives the lithium bromide unit to supply cooling or heating to the buildings.

4.7.2.2 Optimization of Cooling of Internal Combustion Engine

The internal combustion engine is usually cooled by mechanical ventilation, i.e. each internal combustion engine is equipped with air cooling equipment. Given the comprehensive system utilization and site requirements, the project applies a common cooling system for internal combustion engine, lithium bromide unit and centrifugal electrical cooling equipment where several cooling towers are erected at the top of buildings and the cooling water manifolds are introduced the internal combustion engines for cooling. Such method improves the utilization rate of cooling water and reduces the investment in system.

4.7.2.3 Optimization of Reliability of Cooling System

The project has a huge amount of cooling water. Given the safety and reliability of the recycling cooling water system and the pipeline arrangement, the recycling cooling water system is designed as three independent systems. The first independent system is mainly to provide cooling water to four centrifugal chillers, the second system to provide cooling water to three centrifugal chillers while the third system for the two exhaust lithium bromide units and two internal combustion engines that are equipped with two sets of cooling towers and three recycling water pumps. The first and second independent systems may be combined. When operating at low load, the two subsystems are mutually independent while operating at high load the butterfly valve on the manifold is turned on to combine the two subsystems into a bigger independent system, which significantly improve the flexibility of system operation.

In the three independent systems of recycling cooling water, the cooling towers are provided on the roof of higher buildings with nominal height of 42 m. The recycling water pumps are arranged at the level of minus 18 m with no water pool. The recycling cooling water system is unique in China because of its larger amount of recycling cooling water, larger diameter of cooling water pipes and larger height difference (about 70 m) between the inlet pipes of cooling tower and outlet pipes of recycling pumps. To guarantee the safety and reliability of recycling cooling water system, the bidding documents of cooling towers require deeper design of water tray and before engineering design a full and complete investigation has been conducted on the systems of similar types.

4.7.2.4 Energy Saving Technology of Pumps

In this project, the Central Enterprise Headquarters at World Expo Zone B has large volatility of cooling and heating loads so that the cost efficiency of the project is significantly influenced by the energy consumption of pumps. Therefore, the three separate systems of recycling cooling water are equipped with three variable frequency pumps. In addition to energy conservation, they can effectively avoid the vortex caused by excessive start flow where the air inlet in pipes may hamper the normal operation of recycling pumps. The primary hot water pump of boiler and the primary hot water pump of lithium bromide are the pumps of constant frequency while the secondary hot water pump, secondary refrigerated water pump and primary refrigerated pump of lithium bromide are all the pumps of variable frequency. Such combination is more adaptive to load change and improves the comprehensive efficiency.

4.7.2.5 Chilled Water Storage Technology

A modern power grid features the peak and valley difference of the power grid. The chilled storage system is able to shift the peak power consumption of the grid and balance such peak and valley gap. The function of peak shifting and valley filling of the chilled storage system improves the safe operation performance of power grid, improves the efficiency of existing power generation, transmission, distribution and transformation equipment, reduces the transmission and distribution loss and thus cuts down the operation cost of power generation and distribution. It makes full use of non-recyclable resource and contributes significantly to the social and economic benefits. Therefore, the government strongly encourages the power consumption in the valley period. The chilled storage technology can greatly reduce the electric tariff as well as the operating cost.

The project applies single-tank chilled storage technology that the chilled storage and thermal storage of water uses the same tank to reduce the configuration capacity of primary machine system so that its initial investment is lower than conventional air-conditioning system. In chilled storage, outlet water temperature of the chilled storage water system is only 3℃ lower than the conventional system with limited COP reduction of primary machine. Given the energy conservation performance of the whole system (e. g. the temperature is low in nighttime with higher chilling efficiency), the chilled storage system adds almost no extra power consumption. Most systems may even save electric power consumption, i. e. both cost saving and energy saving. Moreover, the chilled storage system has no phase transition and is featured with simple operation, easy maintenance plus lower operation and maintenance costs.

4. 7. 2. 6 Guaranteed Reliability of Power Supply and Uninterrupted Supply of Both Cooling and Heating Load throughout the Year

Since the Central Enterprise Headquarters office building at World Expo Zone B is mainly used for high-end business and office affairs, the power supply system should concurrently supply uninterrupted cold and hot water throughout the year. By quadruple pipelines mechanism, the heat supply load in summer is directly provided by the cooling water of cylinder sleeve of internal combustion engine; the chilled load in winter is directly supplied by the refrigerating water pipes. By using PIPENET hydraulic software and Excel hydraulic calculation model, we take full consideration of changes as per different pipe diameters and velocity so as to guarantee the reliability of water supply to users.

Through the interpretation of system optimization design program on the basis of load analysis results, we make analytic simulation of operation strategy of year-round conditions by different systems and seasons so as to accurately analyze the operational statistics throughout the year.

4. 7. 2. 7 Load Simulation for Higher Power Supply Reliability

The precise load estimation can effectively guarantee the design reliability. We use professional HDY software for specific modeling analysis of buildings at Zone B to simulate the dynamic cooling, heating and power data of all buildings in 8,760 h of a year, which offers the foundation to the unit type selection and operation strategy analysis, effectively saves the initial investment of project and ensures the reliability in the stage of project operation.

A design model is established for the Central Enterprise Headquarters project at World Expo Zone B to calculate the design days and yearly load of air conditioning of the assumed building model. We compare the initial investments of independent type and centralized type of CCHP systems and analyze and calculate the year-round operation energy consumption under the independent CCHP mode so as to finalize the reasonable unit price of heating and cooling. The design is based on the *Rules for Design of Heating Ventilation and Air Conditioning* (GB 50019—2003), *Standards on Energy*

Conservation Design of Public Buildings (GB 50189—2005), *National Technical Measures for Design of Civil Engineering Project-Heating Ventilation Air Conditioning and Power* (GB 50189—2005), etc. Meanwhile in the building energy consumption analysis and energy saving analysis, the influencing factors like heating ventilation air conditioner, meteorological parameters, atmosphere parameters and soil temperature are taken into account.

Through data analysis, provided that the energy supply reliability is guaranteed, the price evaluation system is established to offer basis for the determination of various economic indicators. The analysis serves to:

(1) Determine the installed configurations of cooling and heating source systems of air conditioning as per the design daily load that include single-unit independent thermal source and single-unit heat exchange of centralized supply.

(2) Estimate the respective investment prices of cooling and heating source systems (including the price estimation of electrical, fuel gas, automatic control, civil works, etc. but the air conditioning terminal system is excluded) in accordance with the installed configuration of system.

(3) On the basis of system installed configuration and year-round load, determine the consumption of electric power, fuel gas and water of air conditioning operation in summer and winter with the expense of energy consumption, and then calculate the unit price of energy consumption of thermal operation.

(4) On the basis of investment price, energy expense and annual load, properly consider the maintenance expense and then calculate the unit prices of cooling and heating for contractual energy management.

4.7.3 Project Characteristics

4.7.3.1 Briefing

To make full use of energy, the project uses the CCHP to provide basic cooling and heating load. The insufficient part is provided by the cooling (heating) storage devices, electric refrigerating units and gas boiler.

The following is the major sub-systems of the project:

(1) Combustion system. The heat source necessary for the project is municipal natural gas. The main gas pipe to the prime motor room is divided into two ways, one for internal combustion engine and the other for gas boiler. The mixture of natural gas and air enters the internal combustion engine and produces a large amount of high-temperature fume, which enters the hot water lithium bromide units through generator units. The hot water lithium bromide units utilize the waste heat of smoke and the jacket water of internal combustion engine to obtain heating and cooling. After cooling down of lithium bromide units, the smoke is emitted to the air through smoke silencer and other equipment. The gas-fired hot water boiler is equipped with economizer at the smoke outlet, which cools

down the fume and then emits it to the air. To reduce the heat loss, the smoke outlet of lithium bromide units is equipped with smoke exchanger, so as to further cool down the smoke before emission into the air.

(2) Thermal dynamic system. The water and lubricating oil of internal combustion engine are cooled through high-temperature plate heat exchanger and low-temperature plate heat exchanger. To enhance the utilization of fuels, lithium bromide units is equipped with water-water plate heat exchanger, which uses the cooling water of internal combustion engine to absorb energy for the cooling in summer and heating in winter. After cooling down, the jacket water will then run back to the jacket to realize the circulating cooling for the air cylinder of internal combustion engine.

(3) Cooling system. The cooling system is composed of flue gas hot water lithium bromide system, electrical cooling system and chilled water storage system. There are two lithium bromide units, 3,894 kW for each. There are seven electrical cooling units, with the cooling capacity of 5,977 kW for each. The chilled water storage system is configured with two chilled-water storage tanks (4,100 m^3 + 2,170 m^3) and two public water storage tanks for cooling and heating, with the volume of 500 m^3 each. 367℃ high-temperature flue gas emitted from the power generation of internal combustion engine enters the high temperature generator of flue gas hot water lithium bromide units for cooling cycle. The cooling-down flue gas enters the smoke tube heat exchanger, to heat the cooling water of jacket to 95℃. The heated jacket cooling water enters the lithium bromide units for cooling.

Seven cooling units are provided with the cooling capacity of 5,977 kW each and 41,839 kW in total. The load of single cooling unit is about 10.5% of the design daily load (56.8 MW) and the cooling machine could be adjusted within 25% to 100%. Therefore, the current unit configuration basically meets the cooling supply demand of the project cooling areas at different loads in different seasons.

The heat of cooling storage system comes from centrifugal cooling units. It utilizes the cooling storage at valley period of nights to provide cooling at load peak and valley of day time, to save cost and shave peak load.

(4) Heating system. The heating system is composed of flue gas heat water lithium bromide, boiler heating system and water heat storage system. There are two sets of lithium bromide units, with the heating capacity of 3,000 kW each. There are three natural gas hot water boilers with the heating capacity of 7,000 kW each. The water heat storage devices are two 500 m^3 public water tanks for heat and cool storage.

Flue gas hot water lithium bromide system: Upon power generation, the internal combustion engine emits heat smoke of 367℃, which enters the high-temperature generator of flue gas hot water lithium bromide units for heating cycle. The cooled-down flue gas enters the gas pipe heat exchanger, to heat 95℃ lithium bromide unit storage deer lithium bromide units for further heating.

Upon heat supply, the 60℃ low temperature heat water is pressurized by feedwater pump and then fed into the cylinder block of the boiler. The 90℃ heat water is generated by heat exchange with the high temperature flue gas coil of the boiler. Three ways of 90℃ high temperature heat water enter the main heating pipe to mix with heat water from lithium bromide units and heat storage water tank, and then are pressured and sent to each user through pipelines.

The heat energy of heat storage system all comes from the lithium bromide units, and stores the surplus heat energy produced by the lithium bromide units at the low load of users and provides heat at the load peak of users, to relieve the load pressure of gas boiler.

(5) Circulating cooling water system. The system is mainly used to supply cooling water to internal combustion engine, natural gas hot water type lithium bromide and centrifugal chillers. The cooling tower is responsible for cooling the circulating water. The cooling tower is designed to cool 646 m^3 circulating water per hour in winter, 9,336 m^3 per hour in summer and 5,822 m^3 per hour in spring and fall. The total water volume is 9400 m^3/h. The circulating cooling water system is designed to be divided into three independent systems, to meet different operating conditions.

(6) Electrical system. The project is equipped with two 4 MW generators with all electric energy to be connected to the grid. The output voltage of generators is 10.5 kV. The voltage bus is provided for the generators. 10 kV lines adopts sectionalized double-bus configuration (two sectionalized switch cabinet). The generators are accessed with the voltage bus, among which #1 generator is accessed with Bus Ⅰ while #2 generator with Bus Ⅱ. Two-circuit 10 kV line is accessed with the 10 kV bus of the power system through isolation transformer.

(7) Instrument and control system. The overall structure of the control system includes gas engine, flue gas hot water type lithium bromide units, gas-fired hot water boiler, centrifugal chillers, cool (heat) storage devices and the supervision and operation of start, stop and normal operation of corresponding supporting facilities.

DCS system adopted has the following functions: data acquisition system (DAS), modulation control system (MCS), sequence control system (SCS), furnace safety supervision system (FSSS), electrical control function (including generator-transformer units, station service equipment), industrial TV monitoring system, fire alarm monitoring system and interlocking protection functions. I/O signals are directly accessed to the DCS system in a way of hard wiring, to realize the control and monitoring functions of the whole control system at the DCS operator station. All data acquisition, closed loop control, interlock protection and logic sequential control are all completed by DCS system.

4.7.3.2 Benefits of Multi-energy Mix

The project has CCHP system that combines power generation, cooling in summer and heating in winter to meet the energy demand of Central Enterprise Headquarters of

World Expo Zone B. As the office buildings in this area target at high-end business and office affairs with the demand of heating in summer and cooling in winter, this project adopts quadruple pipe mechanism to realize anti-season power supply, i. e. heating in summer and chilling in winter to improve the energy quality for the central enterprises. The heat supply load in summer is directly provided by the cooling water of cylinder sleeve of internal combustion engine; the chilled load in winter is directly supplied by the refrigerating water pipes. When the temperature is water supply is high, the electrical cooling air conditions should be turned on to cool the water to the required cooling supply parameters for system use. The refrigerating pumps of energy center should ensure the water supply pipe pressure in compliance with the user demand. The year-round load curve is shown in Figure 4. 7. 2 below.

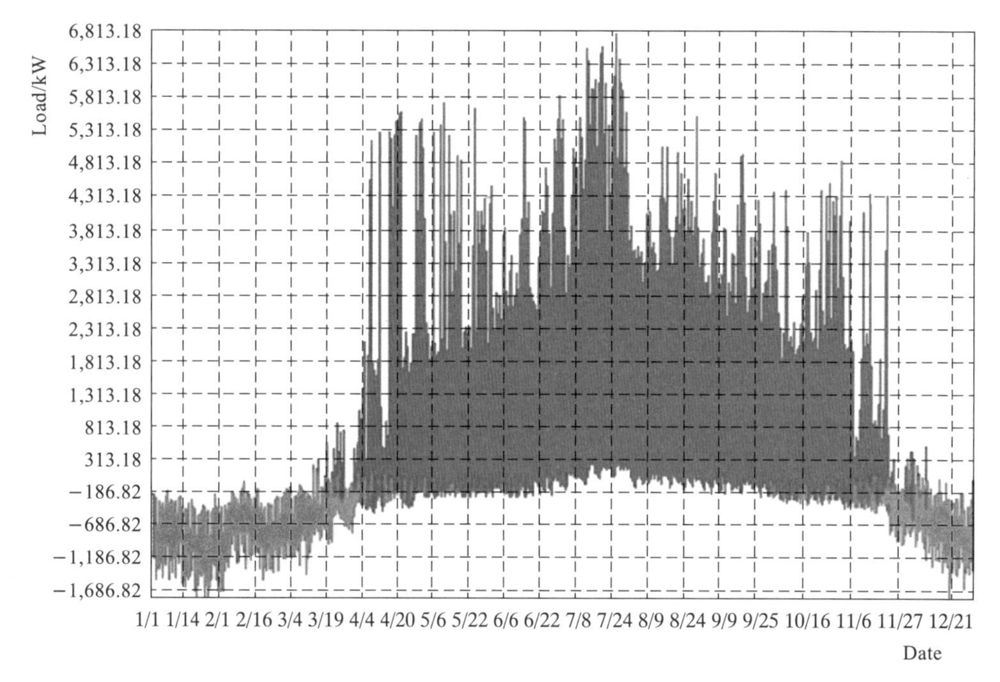

Figure 4. 7. 2 Sketch of Load Curve

Calculated on the all-year hourly load, the accumulative heat consumption in this area throughput the year is 24. 93 GWh and the accumulative cooling consumption is 59. 32 GWh. Thanks to the excellent complementation, the energy center improves the utilization hours of simultaneous cooling, heating and power supply to over 3,300 h with the overall efficiency above 80%. It accelerates the investment return of energy center, and significantly reduces the emission of hazardous gas with higher overall cost efficiency and more environmental benefits.

4. 7. 3. 3 Operation

The aforesaid design optimization leads to satisfactory result of on-site implementation that all processes and schedules are in smooth progress. Figure 4.7.3 is the picture of on-site implementation result.

Figure 4.7.3 The Picture of On-site Implementation Result

By using the aforesaid innovative technology and design methodology, the completed project can supply 210,000 GJ of cooling, 90,000 GJ of heating and 11.47 GWh of electric power. Compared with conventional power supply, it can save 3,611 tons of standard coal each year and reduce the CO_2 emission of 9,269 tons, SO_2 emission of 72 tons. With an increase of energy saving rate of 27% and gross efficiency high up to 85.14%, it is the first underground large-capacity energy center in Shanghai that guarantees all-time stable power supply to the 28 headquarters of central enterprises located at World Expo Zone B. In accordance with the survey result, the CCHP projects of similar type have average all-year full load operation hours about 2,000 h to 3,000 h and the general efficiency is around 80% (e.g. 2,880 h of all-year operation hours and general efficiency of 69.7% in the Central Hospital of Minhang District; 3,500 h of all-year operation hours and general efficiency of 79.8% of Okura Garden Hotel Shanghai). The majority of project is above the ground surface and part of the units is idle, which is incompatible with the present load. Through comparison we can see the indicators of this project are better than the average level in China, which is featured with reasonable unit configuration, underground arrangement, complicated system and better overall performance.

4.7.4 Digitalization and Informatization

This complicated project involves limited underground space, crossover sectors and numerous external conditions. Therefore in the project design, we use 3D design software as per the project characteristics with the modeling of equipment and pipelines of different sectors and the collision simulation to improve engineering safety. In the meantime, in the stage of crossover data provision and engineering, the company's remote operation platform is used to realize all engineering design on the same platform, which avoids the repetitive data provision and revision and effectively reduces design error. The 3D design effect is shown in Figure 4.7.4.

Figure 4.7.4 Implementation Effect of 3D Design Software

4.7.5 Policy Support for Natural Gas CCHP

The natural gas CCHP in China entered the research and development stage in the late 20th century. In the 21st century, it obtained huge development, and is one of key directions for the development of energy structure in the future. The Chinese government attaches great importance to the natural gas CCHP and has offered strong policy support.

4.7.5.1 National Support Policies

(1) Macro energy plan. As early as in 2000, Several Provisions on Developing Combined Heat and Power clearly point out, "each place shall formulate relevant measures for developing and promoting combined heat and power centralized heating according to local conditions." According to the 13th FYP Development Plan for Energy Development, China shall strengthen the overall planning and integration construction of terminal energy supply system. In the new energy consumption areas, it's necessary to implement terminal

integration energy supply projects, promote natural gas CCHP, distributed renewable energy power generation, thermal energy heating and cooling and other energy supply mode land reinforce the energy production coupled integration and complementary utilization of heating, power, cooling and gas. As one of the most important terminal energy supply form, the natural gas CCHP is significant in the national energy plan.

(2) Industry and technical standard. In 2010, China released Technical Specification for *Gas-fired Combined Cooling, Heating and Power Engineering*. This industrial standard set requirements on the system plan, equipment, construction, operation management, environmental protection and so on, which has accelerated the standardization of the industry and provided guarantee for the quality of the natural gas CCHP projects.

(3) Project demonstration promotion. In 2012, China issued the Notice on Releasing the 1st Group of State Natural Gas Distributed Energy Demonstration Projects to select four projects in Beijing, Tianjin, Hubei and Jiangsu as the first group of state natural gas CCHP projects, setting up the standard and providing conditions for the large-scale development of the similar projects.

(4) Economic subsidy policy. The Guide Opinions on Developing Natural Gas Distributed Energy issued in 2011 specifies that "each province, region, municipality and key city shall study and release specific supporting policies according to local conditions and provide natural gas distributed energy projects with certain investment incentives or discount, When the price of distributed energy is determined, the peak shaving of natural gas distributed energy shall be reflected and price concession shall be provided." Using economic subsidy policies to improve the economic advantages of natural gas CCHP projects has actively promoted the industrial development and drive the technical progress of the industry.

4.7.5.2 Local Support Policies

At present, Shanghai, Changsha, Qingdao and other cities in China have released relevant support policies on promoting natural gas CCHP. Taking Shanghai for example:

In 2004, Shanghai issued Opinions on Encouraging the Development of Gas-fired Air Conditioner and Distributed Energy Supply System, proposing to offer subsidy to natural gas CCHP projects as per RMB 700 yuan/kW, and support these projects in the aspects of natural gas price, grid access, preferential taxation for equipment. In 2005, Shanghai issued Technical Specification of Engineering for Distributed Energy System (trial implementation), further setting the locality-based industrial standards based on the national standards. In 2008, Shanghai released Special Support Measures for the Development of Natural Gas Distributed Energy System and Gas-fired Air Conditioner, to increase the subsidy of natural gas CCHP; the equipment subsidy is up to RMB 1,000 yuan/kW. With the updating of subsidy policies, besides equipment subsidy, Shanghai offers another RMB 2,000 yuan/kW to the natural gas CCHP energy supply projects with the average annual energy comprehensive utilization up to 70% and annual utilization hours of 2,000 or above and offer another RMB 500 yuan/kW to those projects with the average annual energy comprehensive utilization up to 80% and annual utilization hours of 3,000 or above. The subsidy for each project is capped at RMB 50 million yuan. Each support policy has played an important role in the vigorous development of natural gas CCHP projects. At present, Shanghai has built multiple natural gas CCHP energy supply stations.

4.7.6 Feasibility of Natural Gas CCHP Application in ASEAN

Natural gas is one of the most important energy resources in ASEAN. Many countries such as Indonesia and Brunei are abundant in natural gas. At present, ASEAN countries are undergoing rapid urbanization and industrialization, and energy supply therefore is of great significance. The natural gas CCHP is suitable for many places in Table 4.7.1 below.

Table 4.7.1　Suitable Places of Natural Gas CCHP

Category	Place
Public Service	Government agencies and institutions
	Cultural, sports, medical, education, transport hubs and other important public facilities
Commercial	Various types of enterprises, industrial parks, economic development zones
	Commercial facilities such as shopping malls, office buildings and hotels
Life of Residents	Resident living area or independent building

The main advantages that the project brings to the region are:

(1) Improve energy efficiency and save fossil energy.

(2) Reduce environmental pollution.

(3) Better project economy, reduce energy consumption costs of users and encourage the development of industry and commerce.

(4) Provide energy consumption protection for government agencies and important public facilities, and enhance the government's public service capabilities.

(5) Utilize underground space for development because of small floor area and promote land intensive management.

4.8 Largest Micro Energy Utilization Project in China— "Six-in-One" Multi-Distributed Renewable Energy Hybrid Project of GCL Energy Center

4.8.1 Project Overview

The "six-in-one" multiple distributed renewable energy hybrid project is built and implemented with GCL Energy Center, the Headquarters of GCL in China as its carrier.

GCL Energy Center is the R&D and office base of GCL New Energy Holdings Limited in Suzhou Industry Park. Located at south of Xincheng Road, west of Changyang Street, north of Xinqing Road and east of Nansan Road, the office, experiment and R&D buildings will be built in two phases where Phase Ⅰ is an experimental building with the total floor area of 20,652 m^2 and the capacity building area of 19,515 m^2 and Phase Ⅱ includes a science and technology building, science research auxiliary building and future energy pavilion with the total floor area of 64,133 m^2. See Figure 4.8.1 and Figure 4.8.2.

Figure 4.8.1 Phase Ⅰ and Phase Ⅱ of GCL Energy Center

Figure 4.8.2 Future Energy Pavilion

GCL Energy Center mainly aims to meet such demands as power, air conditioner cooling, heating and domestic hot water. The design load for its total energy demand is 3,200 kW; the energy saving design at the demand side could save about 2,000 kW compared with the conventional energy configuration. Through implementing the "six-in-one" distributed renewable energy hybrid projects, Phase Ⅰ and Phase Ⅱ PV power generation systems could provide 808 kW electric energy and natural gas distributed CCHP system could provide 400 kW electric energy or 400 kW heat/cool energy. Phase Ⅱ Group Source Heat Pump could provide 5,000 kW (equal to 980 kW electric energy) heat/cool energy. The low wind power generation system could provide 60 kW electric energy. It is also equipped with a 200 kW energy storage system for adjusting the stability of the whole micro energy network. The self supply rate of the Energy Center exceeds 50% (the insufficient part is supplemented by the municipal power supply), and the whole building could save above 30% of energy.

"Six-in-one" multiple distributed renewable energy hybrid project is optimized at both the supply and demand sides with multiple points distributed in all areas covered by the Energy Center. The Energy Center Phase Ⅰ: natural gas CCHP+PV power generation+ energy storage+geothermal heat pump+wind-solar powered road lamp+low wind power gene ration+LED light source.

The Energy Center Phase Ⅱ: roof top solar PV+LED light source+geothermal heat pump+wind-solar powered road lamp.

4.8.2 Key Technology of "Six" Subsystems

The core of GCL Micro Grid lies in the comprehensive utilization of many energy sources of different varieties and gives full consideration to the characteristics of different energy sources, such as wind energy and solar energy. These energy sources are organically integrated to an energy microgrid to collaboratively optimize multiple energy sources and at the same time provide users with different varieties of energy such as power, cool, heat and gas, to meet the multiple energy demands of users. In this way, it has increased the efficiency of comprehensive energy utilization, reduced the total cost of energy consumption and realized the goal of energy conservation, emission reduction and environmental protection.

The "six-in-one" multi-renewable energy hybrid project is composed of energy saving systems such as natural gas distributed energy system, PV power generation system, low wind power generation system, energy storage system and LED light source as well as microgrid monitoring system. The project has achieved technological breakthrough of highly integrated operation in multiple energy forms while completing the core technologies of each subsystem. The project has become an example of China's micro energy grid projects for the subsequent application and development.

4.8.2.1 Natural Gas Distributed Energy System

The 400 kW-level natural gas CCHP energy station is built for the natural gas distributed energy system, which is equipped with one 400 kW gas-fueled internal combustion engine plus one exhaust and hot water lithium bromide absorption refrigeration unit (refrigerating capacity of 400 kW). By CCHP technology, it provides the energy center with 400 kW electric energy and 400 kW cooling/heating energy. See Figure 4.8.3.

Figure 4.8.3 Natural Gas Distributed Energy Station

4.8.2.2 PV Power Generation System

The PV power generation system uses the solar battery board made by the PV effect to produce electric energy. In this project, the PV power generation systems are installed on the roofs of office building and garage. See Figure 4.8.4, Figure 4.8.5 and Figure 4.8.6.

Figure 4.8.4 Roof Uniaxial PV Power Generation

Figure 4. 8. 5 Panorama of Roof PV

Figure 4. 8. 6 Garage PV

4. 8. 2. 3 Low Wind Power Generation System

It fully utilizes the wind energy of the energy center area, which generates power when wind speed reaches 2. 8 m/s. Once completed, the installed capacity of low wind power generation system is 60 kW. See Figure 4. 8. 7 and Figure 4. 8. 8.

4. 8. 2. 4 Energy Storage System

As part of the microgrid system, the energy storage system consists of lithium iron phosphate batteries, ESMU, ESBMS, system group end control and ESGU with capacity of 200 kWh. See Figure 4. 8. 9.

4. 8. 2. 5 Energy Conservation System like LED Source

The project uses GCL-branded LED lights of 135 kW and wind-solar-storage powered

Figure 4. 8. 7 Horizontal Axial Low Wind Power Generation (50 kW)

Figure 4. 8. 8 Vertical Axial Breeze Power Generation (10 kW)

road lamps of 4 kW that can save 30% of power consumption compared with ordinary light sources. See Figure 4. 8. 10.

4. 8. 2. 6 Microgrid Monitoring System

The establishment of a microgrid monitoring system realizes multiple dispatching advancements, mainly as follows: it gives priority to low-carbon dispatching, PV and wind power, and store the excessive energy; the dispatching at the supply side corresponds to the demand side, to reduce peak shaving at load side; it dispatches energy efficiency and

Figure 4.8.9　200 kW Energy Storage System

Figure 4.8.10　GCL-branded LED Light Sources

decides electric power by heat; the research on the capability of disturbance resistance of microgrid offers experience for demonstration and wide application.

In the design and implementation of dispatching and control strategy for the micro energy network of the energy center, the following aspects should be considered:

(1) Dispatching control of multi power sources. The microgrid system of the energy center is the grid-connected power supply system, which is connected with the grid but does not feed in power to the grid. The power generated by the system is consumed by itself. The distributed energy of energy center has diversified types including an internal combustion engine CCHP system, solar energy, wind power and battery storage system.

Considering different exterior conditions, the microgrid has three operation modes namely grid-connected operation, isolated island operation and mode shift.

When in the grid-connected mode, the microgrid connection with the power distribution network should comply with the interface requirements of the network. Hence, the grid-connected microgrid integrated with natural gas distributed power, PV, low wind power and energy storage , can reduce the grid energy consumption and make the regional load more reliable with no impairing electric energy quality.

When in isolated island mode, the microgrid is able to sustain its own voltage and frequency. As a large number of electronic and electrical equipment is used as interfaces in microgrid, their small or no inertia, poor overload capacity, intermittent output energy of distributed power source of renewable energy and the volatility of load power increase the difficulty of control of microgrid frequency and voltage.

When microgrid shifts between these two modes, it is critical to maintain its stability. As the microgrid absorbs or feeds out power to the grid in grid-connected operation, if the microgrid suddenly shifts from the grid-connected mode to the isolated island mode, the imbalance between the electric energy and load demand may lead to system instability. Therefore, it is very important to adopt appropriate mode shift control method.

The primary functions of multi energy dispatching system include: coordinative operation of multiple power sources; quick separation in case of failure of municipal grid or microgrid; relay protection of the isolated island operation of microgrid; regulation and control of all distributed power sources connected with microgrid; uninterrupted shift of operation modes; balance between power source and load in the isolated island mode of microgrid; reactive power mediation of power source; energy management and economical operation of microgrid.

(2) Dispatching control of multi thermal sources.

1) The main purpose of GCL Energy Center is to provide office and R&D workplaces to the managerial staff of the group, including both electric energy consumption and cooling and heating energy demand, where the latter is used for the heating ventilation system and canteen hot water of central air conditioning. The cooling and heating sources mainly include gas-fueled boiler, internal combustion engine+exhaust aluminum bromide unit, chilled water unit and geothermal heat pump.

2) The main purposes of multi heat and energy sources dispatching system are to realize the management, coordination and economical operation of multi heat and energy sources, realize the compatibility of cooling/heating load demand with energy supply, and realize the physical cascade connection of all heat and energy sources with priority given to the renewable energy.

(3) Principle for dispatching techniques.

1) The low wind and solar power generation has no exhaust emission, consumption

of fossil fuel or pollution with lower operation cost. However, the output power is decided by wind and solar energy, which is inconsistent and thus not fully incompatible with load. The batteries or other auxiliary systems are required to be the fundamental load of microgrid to store all the power generated.

2) The micro gas-fueled engine is featured with small size, light weight, high power generation efficiency, less pollution, simple operation maintenance and fit for exclusive dispatching. The micro gas-fueled engine with electric and electronic conversion and control interface can follow the voltage and frequency change of power grid, and work for load tracking, peak shaving and valley regulating. In addition to the basic control of active power, it also regulates the reactive power of system output to regulate the voltage and adjust power factor. Therefore, it is presently one of the most mature and commercially competitive distributed power sources.

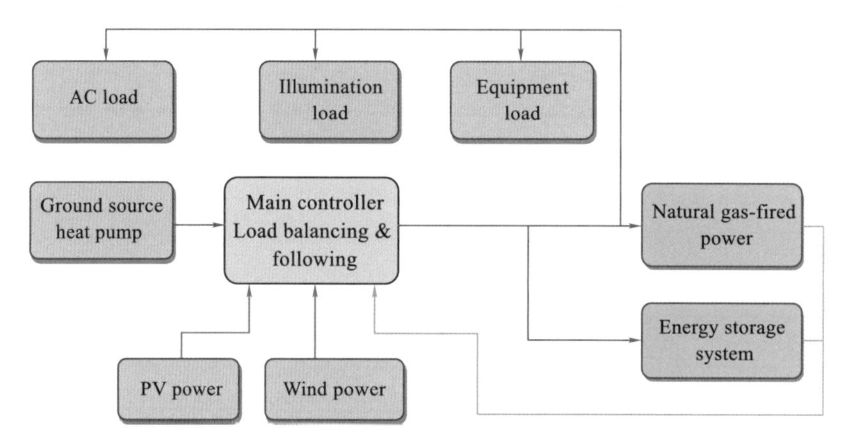

Figure 4. 8. 11 Sketch of Dispatching Control Theory of Micro Energy Grid

The dispatching control of the micro energy grid (see Figure 4. 8. 11) mainly abides by the following basic strategies and principles:

Priority of safety: the dispatching of microgrid should firstly guarantee the safety operation of grid connected system and its own system.

Priority of low carbon: the renewable energy like wind and solar power generation and geothermal heat pump should be prioritized.

Priority of efficiency: under the principle of "electric power decided by heat and cascade utilization", the residual heat resource can be fully utilized to improve the comprehensive efficiency of energy utilization.

Priority of cost efficiency: by different energy financial calculation models, the most economical operation mode is selected.

Priority of demand: in general control the users at demand side are prioritized to guarantee the stability at power source side.

(4) Proportional Configuration of Multi Energy.

As the site of office building, the GCL Energy Center has large variance of load in day and night where the energy consumption proportion in daytime is 5 to 6 times of that in nighttime, and plus the incapable solar power generation at night, the utilization of multi clean energy is needed to reduce energy cost. The proportion of clean energy should not be too big so as to guarantee the full-load operation of all clean renewable energy and the overall efficiency of the whole system. A sensitivity analysis of energy proportion has been conducted as per the roof PV capacity, power and thermal load of energy center and the result suggests the most economical self-sustained proportion around 70%.

(5) Peak regulating energy proportion.

1) The load regulating method of the energy center is the two-way peak regulating for both power supply and demand sides. Hence compared with single-side peak regulating, the capacity of regulating equipment may be halved, which reduces the peak regulating energy proportion. The energy consumption margin and regulating scope are calculated on the basis of energy consumption load of energy center, and the results suggest around 35% of peak regulating capacity at both supply and demand sides.

2) The energy dispatching of the micro energy grid realizes automatic allocation of multi thermal and power sources, and utilizes multi energy in a safe, low-carbon and efficient way through physical setting, parameter monitoring and automatic adjustment. It follows the five principles of "priority of safety, low carbon, efficiency, economy and demand" to achieve energy utilization in a safe, clean, efficient and low-carbon way and improve all-round energy quality.

4.8.3 Project Characteristics

The GCL "Six-in-One" micro energy grid has six characteristics of energy creation, green energy, multi energy, energy storage, energy saving and micro energy.

4.8.3.1 Energy Creation

The renewable energy and curtailed energy are recycled to create serviceable heating, power and cooling energy and optimize the energy transformation model.

4.8.3.2 Multi Energy

The source form and supply form of energy realizes the full and mutually complementary utilization of multiple energy resources to guarantee the safety and reliability of energy usage.

4.8.3.3 Green Energy

All energy used is the energy of zero or trivial emission (solar energy, wind energy and natural gas) that is ecological friendly, green and low-carbon.

Energy saving: it realizes the cascade utilization of energy to significantly reduce energy consumption, applies green energy consumption equipment to improve the energy utilization efficiency, and saves energy in an intelligent way.

4.8.3.4 Micro Energy

The production of energy can be lowered to kW level that offers separate energy solution to each micro user.

The GCL Six-in-One micro energy grid achieves technology breakthrough in four aspects, i. e. peak regulating, low-level thermal energy utilization, energy storage and micro energy grid.

The Six-in-One micro energy grid realizes dimensional hybrid of renewable energy and new energy, gives full play to different forms of energy, provides the grid with diversified energy with guaranteed safety, and solves the four technical problems of energy utilization namely the storage, peak regulating, low-level thermal energy utilization and micro energy grid. The microgrid system and its equipment has strong automation control and variable conditions capabilities and convenient start/stop that quickly starts from zero load to full load. A modern energy supply mode of energy supply in the vicinity of load center is implemented that randomly matches the change of peak-valley difference of load center; the low-level thermal energy technology solves the energy peak regulating problem and applies the geothermal heat pump and hot water power generation technology to make use of the abandoned low temperature (below 80℃) energy to produce high level energy (electric power or steam) for refrigeration, power generation and heating reduction, which shifts the energy from low-temperature heat source to high-temperature heat source.

4.8.4 Socioeconomic Benefits

The research and implementation of micro energy grids aims to realize the diverse, efficient and environmental friendly energy allocation within certain scope, and the optimal proportion of local consumption of energy and self-sufficiency. It will greatly expand the application space of clean energy at each small unit nearest to the demand side and be of practical and promotional significance to areas not covered by the national grids, such islands and frontier areas, while solving the impact on grid stability, energy loss and dispatching pressure of clean energy in the long-distance transmission.

GCL Six-in-One Micro Energy Grid project (see Figure 4.8.12) has a total investment of around 40 million CNY. The whole industry chain of microgrid is relatively costly. Therefore, the Central and local governments of China allow to build the trial operation and demonstration projects within the scope of mature conditions and equipment, and grant 10% to 20% installed capacity subsidy. In addition, China will offer extra RMB 0.42 yuan/kWh for power generated by the PV system to boost the prosperity of microgrid industry chain. Meanwhile, enterprises in charge of implementing these projects are also constantly studying and exploring the technological progress and cost reduction space for the microgrid projects and developing innovative profit models, to promote the sustainable and healthy development of the industry.

Figure 4. 8. 12　Real Scene of GCL Energy Center

4. 9　Other Cases—National Demonstration Project of Wind-Solar Hybrid Power Generation, Storage and Transmission

4. 9. 1　Project Overview

Wind power, PV power generation and other new energy sources are characterized by fluctuating, intermittent and stochastic and challenged by predication, difficult dispatching and control and other technical problems, which has limited the centralized development and transmission. To solve these problems and promote the sustainable development of new energy in China, without reference of any engineering construction precedence, SGCC follows the leading principle of all-round innovation and breakthroughs in each field and proposes the world's first wind-solar-storage-transmission technical route based on the national conditions and has built and put into operation the national wind-solar-storage-transmission demonstration project (the "Demonstration Project"). Located at Zhangbei County, Hebei Province, the Demonstration Project is the world largest new energy comprehensive demonstration project with the highest comprehensive utilization level. Combining the "wind power, solar power, energy storage system and smart power transmission", it is the pioneer showing the advanced and innovative wind, solar, storage and transmission combined power system.

The demonstration project (see Figure 4. 9. 1) was planned to build 500 MW wind power, 100 MW solar power and 70 MW energy storage facilities: Phase Ⅰ project of 100 MW wind power, 40 MW solar power and 20 MW energy storage was completed and

Figure 4. 9. 1 Panorama View of the Demonstration Project

commissioned on December 25, 2011. Phase Ⅱ project of planned 400 MW wind power, 60 MW solar power and 50 MW energy storage was commissioned for full-capacity power generation by the end of 2014 except the energy storage and heavy-duty wind turbines.

The Demonstration Project applies the world first technical route of wind-solar power, energy storage and transmission combined power generation. Through studies on a hybrid system of wind-power-storage power generation and output and optimal configuration, integrated technologies for multiple battery energy storage power plant, combined power generation and panoramic monitoring and coordination control, it has successfully realizes the smooth output of new energy power, which offers a solution to the world-class difficulty of scale-up development and utilization of new energy and enables the clean energy to the most probable power form for substituting thermal power.

4. 9. 2 Key Technology

The key technology innovation includes:

(1) Developing the first large-scale wind and solar storage application model, comprehensively revealing the complementary rules of wind and solar power generation and storage output and the coupling mechanism with the sending power grid; establishing a large-scale wind and solar storage theoretical model to support the construction and demonstration operation of the national wind and solar energy storage demonstration project, creating a new integrated development in China.

1) Establishing a hybrid model of energy storage and wind power output involving multiple application targets, multiple time scales, and multiple demand responses.

2) Establishing a theoretical model for the coupling of wind and solar energy storage systems and grids for the first time.

3) Developing the capacity optimization and allocation software for wind and solar energy storage combined power generation systems for the first time with the evidence of demonstration projects.

(2) Overcoming the key technologies of active wind-solar-storage grid-connected control, proposed, proposed active and reactive coordination control methods, established a dispatching model for wind-solar power generation convergence area, independently developed 100-megawatt-level wind and solar storage panoramic monitoring system, realizing friendly interaction between new energy generation and grid dispatching and increasing the acceptance capacity of the grid for new energy.

1) Breaking through the bottleneck of information interaction between various types of equipment in wind-solar-storage combined power generation, established a unified information model and communication protocol, and built an integrated high-speed real-time information fusion platform.

2) Realizing the visualized panoramic monitoring and integrated control of wind farms, photovoltaic power plants, energy storage power plants, and smart substations for the first time.

3) Putting forward the multi-configuration timing output optimization and multi-source reactive voltage auto-coordinated control method for the first time to achieve functions such as power smoothing, tracking planning, peak shaving and frequency modulation, and voltage control, and achieve the power quality close to that of a conventional power supply.

(3) Making breakthrough in large-capacity battery grouping and large-scale battery energy storage integration technology, independently developed a layered coupled real-time control system, and built the world's largest multi-type energy storage power plant with the core performance parameters such as control accuracy, energy conversion efficiency and response speed reaching the international level.

1) Proposed large-capacity grouping technology and battery system cascading technology involving the dynamic consistency of the battery, to realize the integration of the largest multi-type battery energy storage power plant system in the world, and the energy conversion efficiency is more than 90%.

2) Developed an energy storage monitoring system in which the centralized control and distribution control are coupled to each other so that the overall response time of the energy storage power plant is less than 900 milliseconds and the output error is less than 2%.

3) Developed coordinated control and energy management software for energy storage power plants and enabled multiple advanced application functions for the first time in the

same energy storage plant.

(4) Developed the state assessment technology of large-capacity wind-solar-storage combined power generation system, established a state assessment system based on the dual indicator dimension of reliability and power generation performance, developed the first set of state assessment auxiliary decision-making system for wind-solar-storage power generation equipment, ensured the safe and reliable operation of demonstration power plants and gave play to the leading role of operation and maintenance technology for the national demonstration project.

1) Put forward the equipment condition assessment index method for wind-solar-storage system for the first time and established an efficient evaluation system was established.

2) Developed the first set of state assessment aided decision systems integrating state assessment analysis, intelligent fault diagnosis and maintenance guidance, and power plant production support management.

3) Released series standards covering debugging and acceptance, operation and maintenance and technical supervision.

(5) Integrated the most advanced equipment technologies in China, established the world's largest wind-solar-storage-transmission demonstration project with the highest level of intelligence and flexible operating mode for the first time, successfully demonstrated the comprehensive effect of improving the quality of new energy generation and power grid acceptance, and become a technological achievements display platform, key equipment inspection platform and scientific research incubation platform in the field of new energy power generation.

1) Set up a demonstration platform for the first application of the first set of equipment with the national advanced technology, to centrally demonstrate and compare the application performance of various new types of wind power and photovoltaic equipment and promote the upgrade of the domestic new energy equipment industry.

2) Realized the application of direct-mounted SVG in new energy power plants and the all-weather participation of photovoltaic power plants in power grid reactive power regulation and other applications, greatly improving the dynamic reactive power support capability.

3) Introduced smart substation to participate in the demonstration operation for the first time, fully meeting the application research requirements of different configurations and ratios of wind and solar energy storage.

4.9.3 Project Characteristics

4.9.3.1 Briefing

The demonstration project targets at the "grid friendly" new energy power generation, combines the advanced technology in the sectors of power grid, wind power, solar power, energy storage, etc. and is an advanced, flexible and demonstrative new energy project.

(1) Wind farm. It is completed in two phases that covers the existing domestic capacity of 1.0 MW, 2.0 MW, 3.0 MW and 5.0 MW, the different manufacturers like Goldwind, Xuji and XEMC and 177 wind power units of different types such as double-fed induction generator (DFIG), direct-driven wind turbine and vertical axial wind turbine (VAWT), which is the first grid-connected friendly wind farm in China with the most diversified types and leads the development trend of onshore wind power generators of larger capacity and higher efficiency.

(2) Solar power plant. Solar power plant Phase Ⅰ with the total capacity of 40 MW includes the demonstration zone and the test zone. The demonstration zone uses polysilicon PV modules on fixed frames with a total capacity of 28 MW; the modules in the test zone include polysilicon, monocrystalline silicon, amorphous film, back contact and high magnification types of modules equipped with fixed, tilted uniaxis, flat uniaxis and double-axis tracking systems for the different combination and comparison analysis of various modules and tracking systems. Given the operation result of Phase Ⅰ project and with comprehensive consideration of cost factor, solar power plant Phase Ⅱ all uses polysilicon PV modules on fixed frames.

(3) Energy storage power plant. Energy storage plant Phase Ⅰ covers four types of electrochemical energy storage battery including lithium iron phosphate battery of 14 MW/63 MWh, all vanadium redox flow battery of 2 MW/8 MWh, colloidal lead acid battery of 2 MW/12 MWh and lithium titanate battery of 1 MW/0.5 MWh. For the lithium iron phosphate batteries, the products of manufacturers like BYD and ATL are put into operation in Phase Ⅰ project. Phase Ⅱ energy storage plant plans the cascade utilization of electric motor batteries totaling 9 MW and the demonstrative application of the remaining 41 MW energy storage will be conducted in accordance with the R&D progress of technology and products.

4.9.3.2 Operation

The demonstration project uses the wind-solar-storage combined monitoring system to plan the combined output of wind power, solar power, power storage facility, etc. as a whole to realize the automatic combination state of seven operation modes (wind, solar, energy storage, wind-solar hybrid, wind-storage hybrid, solar-storage hybrid, and wind-solar-storage hybrid). The shifts between these operation modes of wind-solar-storage combined power generation function such as smooth output, tracking plan, peak shaving and valley filling and frequency regulating that to some extent guarantee the foreseeable, controllable and dispatch able new energy power generation.

(1) Smooth output. The power plant takes reference of and improves the passive low pass filter strategy and makes use of the charging and discharging of energy storage system

to guarantee the smooth and stable total output of the wind-solar-storage combined power generation system. Its power change rate is less than the national standard limits and close to the regular power source level where the combined output fluctuates less than 5% within 10 min.

(2) Tracking plan. The wind-solar-storage combined monitoring system automatically adjusts the output in accordance with the power generation plan curve of the day assigned by power grid dispatching organ, and monitors the charging or discharging capacity of energy storage system in a real-time way to maximize the wind and solar output. When the wind and solar output satisfies the generation plan, the residual power is charged to energy storage system, and will be released in case of insufficient wind and solar power output. The deviation of combined output tracking from the plan requirement is less than 3%.

(3) Peak shaving and valley filling. On the basis of wind and solar resource and load demand, the wind-solar-storage combined monitoring system can maintain the operation of energy storage system under charging or discharging status: in nighttime or special time of smaller load, the energy storage system is under charging condition due to larger output of wind and solar power; in daytime of peak consumption load, the energy storage system releases the stored energy for peak-valley shift.

(4) Energy storage and frequency regulating. By information interchange of wind-solar-storage combined monitoring system, the superior dispatching organ knows the real-time status of energy storage system and by judgment of present state of energy storage system, it gives the control order of automatic power generation so that the energy storage system quickly responds to the power demand of system in seconds to participate in system frequency regulation with the error less than 0.5%.

4.9.3.3 Innovations

Supported by many national sci-tech plans and self-innovation, the demonstration project uses the key technology of scale-up development and utilization of new energy.

Firstly it solves the world problem of operational control of wind-solar-storage combined power generation that generally controls the combined output of wind power, solar power and energy storage, makes use of the instant charging and discharging feature of energy storage, and quickly and precisely adjusts the wind and solar output so that the total output of new energy is smooth and controllable and reaches the criteria of regular power source.

Secondly it masters the overall inspection and testing technology competency of wind power, solar power and energy storage equipment, improves the overall equipment level and availability and solves the problem of large-scale wind power tripping off (no large-scale wind power tripping off occurred after 2013).

4.9.4 Socioeconomic Benefits

Since its launch into operation, the project has maintained stable operation and so far it has totally fed out more than 4,178 GWh quality green electric energy. By self innovation and icebreaking technology, the annual available rate of wind power units was uplifted from 93.1% to 95.9% and the PV unit power output per annum was uplifted by 12.85%. By reference to the annual power output of 1,250 GWh under the international common model, the demonstration project saves 420,000 tons of standard coal and reduces CO_2 emission of 900,000 tons each year.

Supported by engineering practice, the demonstration project has made breakthrough in plenty of key technology, among which the capacity proportioning optimization technology, integration technology and operation control technology of large capacity wind-solar-storage plants have been applied in the projects in Qinghai, Gansu, Ningxia, etc. with conspicuous results and effect. The demonstration project takes a lead of scale-up application of energy storage and promotes the booming level of general assembly in China, whose industry scale expands by 40 times and its cost decreases by 78%.

Figure 4.9.2 President of IEC Visits and Gives Guidance to the Demonstration Project

More than 60 nations like US and Germany, 28 international organizations like CIGRE and IEA and more than 1,000 experts visited the demonstration project. The IEC Chairman praised it as "amazing and world leading", "the project of God" (See Figure 4.9.2). It also provides international assistance to 15 nations, which has become a "National Card" to show China's latest achievement in the new energy development to the world.

4.9.5 Future Development

Focusing on the establishment of friendly grid connection standard model of new energy power plant, SGCC will continue developing other demonstration projects like deep chilled gasification compressed air energy storage and simulative synchronous generator, etc. on the basis of engineering of Phase Ⅱ wind-solar-storage-transmission project to amplify its demonstrative and leading effect.

The simulative synchronous generator technology introduced the electromechanical transient equation to the control link of power electronic equipment, enabling new energy to adaptively participate in the frequency and voltage control of the system. Meanwhile, it also has the oscillation function of damped system, enabling the new energy power generation to have external characteristics similar to those of conventional thermal power plant. At present, China has built the national wind-solar-storage-transmission demonstration project including 59 sets of 118 MW wind power and 24 sets of 12 MW PV inverter software and hardware technical transformation and two 5 MW power plant type virtual synchronous generator projects. This action will further enhance the capability of wind power, PV and new energy power plants to actively participate in the primary frequency modulation and voltage regulation of the grid, providing an exemplar standard for the friendly grid connection of new energy and enhancing the safe and stable operation level of the grid. For the next step, SGCC will further promote the application of virtual synchronous generator technology. The cryogenic energy storage demonstration project will be equipped with two 12.5 MW×8 h cryogenic energy storage systems to enhance the capability of new energy power plant to stabilize the intermittent fluctuation of new energy at large time scale. Moreover, it will further promote the cascade utilization of retired batteries from electric motors to lower the lifecycle investment of energy storage system and improve the cost efficiency of energy storage project.

References

[1] Cao Bei. Comprehensive Configuration Optimization of Hybrid Power Microgrid [D]. Nanchang: Information Engineering School of Nanchang University, 2014: 3-4.

[2] Sun Nan, Xing Deshan, Du Hailing. Development and Application of Wind and Solar Hybrid Power Generation System [J]. Shanxi Power, 2010 (4): 55-56.

[3] Yang Sangen. Research on Problems of Wind and Solar Power Combined Dispatching [D]. Chengdu: University of Electronic Science and Technology, 2015: 3-8.

[4] Shu Jie. Research on Independent Microgrid Technology Based on Distributed Renewable Energy Power Generation [D]. Guangzhou: Sun Yat-sen University; 2010.

[5] Zhang Buxiao. Research and Analysis of Combined Operation of Wind-solar Hybrid Power Generation and Pumped Energy Storage [D]. Jinan: Shandong University of Science and Technology, 2014.

[6] Chen Jimei. Initial Experience of Grid-connected Hybrid Operation of Wind Energy Resource and Small Hydropower Plant with Reservoir Regulating Capacity [J]. Wind Power. 1998 (2): 25-28.

[7] Yang Sangen. Research on Problems of Wind and Hydropower Combined Dispatching [D]. Chengdu: University of Electronic Science and Technology, 2015.

[8] Ma Chengfeng. Research on Wind Power Forecast and Wind-Hydro Coordinative Operation [D]. Nanjing: Nanjing Normal University, 2012: 3-4.

[9] Wang Jin, Li Xinran, Yang Hongming, et al. Regional Distributed CCHP Energy System Integration Solution Synchronous with Electric Power System [J]. Automation of Electric Power Systems, 2014 (16): 16-21.

[10] Zhang Boquan, Yang Yimin. Status Quo and Development Trend of Wind and Solar Energy PV Power Generation [J]. China Power, 2006 (6): 65-69.

[11] Zhu Fang, Wang Peihong. Application and Optimization of Wind and Solar PV Hybrid Power Generation [J]. Shanghai Power, 2009 (1): 23-26.

[12] Wang Sheliang, Feng Li, Zhang Ping, et al. Hybrid Energy Promoting New Energy Development [J]. Northwest Hydropower, 2014 (6): 78-82.

[13] ASEAN Center for Energy, China Renewable Energy Engineering Institute. Report on ASEAN Power Cooperation Report (Highlights) [R]. Beijing: CREEI, 2016.

[14] POWERCHINA Northwest Engineering Corporation Limited. Research Report on Promoting Renewable Energy Consumption Based on Hybrid Energy Systems in Northwest Area [R]. Xi'an: POWERCHINA Northwest Engineering Corporation Limited, 2017.

[15] POWERCHINA Northwest Engineering Corporation Limited. Feasibility Study Report of Xinjiang Fukang Pumped Storage Power Plant (Audited Edition) [R]. Xi'an: POWERCHINA Northwest Engineering Corporation Limited, 2015.

[16] POWERCHINA Northwest Engineering Corporation Limited. Special Report of Normal Pool Level Selection in Feasibility Study Phase of Xinjiang Hami Pumped Storage Power Plant [R]. Xi'an: POWERCHINA Northwest Engineering Corporation Limited, 2017.

[17] POWERCHINA Northwest Engineering Corporation Limited. Program of Feed-out Planning of Wind Power Base in Xinjiang Zhundong Region [R]. Xi'an: POWERCHINA Northwest Engineering Corporation Limited, 2015.

参 考 文 献

[1] 曹蓓．多能源互补微网的综合优化配置 [D]．南昌：南昌大学信息工程学院，2014.

[2] 孙楠，邢德山，杜海玲．风光互补发电系统的发展与应用 [J]．山西电力，2010 (4)：55-56.

[3] 杨三根．风电与水电联合调度问题研究 [D]．成都：电子科技大学，2015.

[4] 舒杰．基于分布式可再生能源发电的独立微网技术研究 [D]．广州：中山大学，2010.

[5] 张步晓．风光互补发电与抽水蓄能联合运行研究与分析 [D]．济南：山东科技大学，2014.

[6] 陈集梅．利用风能资源与具有水库调节能力的小水电站并网互补运行的初步体验 [J]．风力发电，1998 (2)：25-28.

[7] 杨三根．风电与水电联合调度问题研究 [D]．成都：电子科技大学，2015.

[8] 马成风．风电预测及风电—水电协调运行的研究 [D]．南京：南京师范大学，2012.

[9] 王进，李欣然，杨洪明，等．与电力系统协同区域型分布式冷热电联供能源系统集成方案 [J]．电力系统自动化，2014 (16)：16-21.

[10] 张伯泉，杨宜民．风力和太阳能光伏发电现状及发展趋势 [J]．中国电力，2006 (6)：65-69.

[11] 朱芳，王培红．风能与太阳能光伏互补发电应用及其优化 [J]．上海电力，2009 (1)：23-26.

[12] 王社亮，冯黎，张娉，等．多能互补促进新能源发展 [J]．西北水电，2014 (6)：78-82.

[13] 东盟能源中心，水电水利规划设计总院．东盟能源电力合作报告（简要本）[R]．北京：水电水利规划设计总院，2016.

[14] 中国电建集团西北勘测设计研究院有限公司．西北地区促进可再生能源消纳多能互补研究报告 [R]．西安：中国电建集团西北勘测设计研究院有限公司，2017.

[15] 中国电建集团西北勘测设计研究院有限公司．新疆阜康抽水蓄能电站可行性研究报告（审定本）[R]．西安：中国电建集团西北勘测设计研究院有限公司，2015.

[16] 中国电建集团西北勘测设计研究院有限公司．新疆哈密抽水蓄能电站可行性研究阶段正常蓄水位选择专题报告（审定本）[R]．西安：中国电建集团西北勘测设计研究院有限公司，2017.

[17] 中国电建集团西北勘测设计研究院有限公司．新疆准东地区风电基地外送规划方案 [R]．西安：中国电建集团西北勘测设计研究院有限公司，2015.